AN INTRODUCTION TO THE THEORY OF REPRODUCING KERNEL HILBERT SPACES

Reproducing kernel Hilbert spaces have developed into an important tool in many areas, especially statistics and machine learning, and they play a valuable role in complex analysis, probability, group representation theory, and the theory of integral operators. This unique text offers a unified overview of the topic, providing detailed examples of applications, as well as covering the fundamental underlying theory, including chapters on interpolation and approximation, Cholesky and Schur operations on kernels, and vector-valued spaces. Self-contained and accessibly written, with exercises at the end of each chapter, this unrivaled treatment of the topic serves as an ideal introduction for graduate students across mathematics, computer science, and engineering, as well as a useful reference for researchers working in functional analysis or its applications.

Vern I. Paulsen held a John and Rebecca Moores Chair in the Department of Mathematics, University of Houston, 1996–2015. He is currently a Professor in the Department of Pure Mathematics at the Institute for Quantum Computing at the University of Waterloo. He is the author of four books, over 100 research articles, and the winner of several teaching awards.

Mrinal Raghupathi is a Lead Quantitative Risk Analyst at USAA. His research involves applications of reproducing kernel Hilbert spaces, risk analysis, and model validation.

CAMBRIDGE STUDIES IN ADVANCED MATHEMATICS

All the titles listed below can be obtained from good booksellers or from Cambridge University Press. For a complete series listing visit: www.cambridge.org/mathematics.

An Introduction to the Theory of Reproducing Kernel Hilbert Spaces

VERN I. PAULSEN

University of Waterloo, Waterloo, Ontario, Canada

MRINAL RAGHUPATHI

CAMBRIDGE
UNIVERSITY PRESS

University Printing House, Cambridge CB2 8BS, United Kingdom

One Liberty Plaza, 20th Floor, New York, NY 10006, USA

477 Williamstown Road, Port Melbourne, VIC 3207, Australia

4843/24, 2nd Floor, Ansari Road, Daryaganj, Delhi - 110002, India

79 Anson Road, #06-04/06, Singapore 079906

Cambridge University Press is part of the University of Cambridge.

It furthers the University's mission by disseminating knowledge in the pursuit of education, learning and research at the highest international levels of excellence.

www.cambridge.org
Information on this title: www.cambridge.org/9781107104099

First published 2016

A catalogue record for this publication is available from the British Library

Library of Congress Cataloging in Publication data
Names: Paulsen, Vern I., 1951– | Raghupathi, Mrinal.
Title: An introduction to the theory of reproducing kernel Hilbert spaces /
Vern I. Paulsen, Mrinal Raghupathi.
Description: Cambridge : Cambridge University Press, 2016. | Series:
Cambridge studies in advanced mathematics ; 152
Identifiers: LCCN 2015042446 | ISBN 9781107104099
Subjects: LCSH: Hilbert space–Textbooks. | Kernel functions–Textbooks.
Classification: LCC QA322.4 .P38 2016 | DDC 515/.733–dc23
LC record available at http://lccn.loc.gov/2015042446

ISBN 978-1-107-10409-9 Hardback

Contents

Preface

This book grew out of notes for seminars and courses on reproducing kernel Hilbert spaces taught at the University of Houston beginning in the 1990s.

The study of reproducing kernel Hilbert spaces grew out of work on integral operators by J. Mercer in 1909 and S. Bergman's [4] work in complex analysis on various domains. It was an idea that quickly grew and found applications in many areas.

The basic theory of reproducing kernel Hilbert spaces (RKHS) goes back to the seminal paper of Aronszajn [2]. In his paper, Aronszajn laid out the fundamental results in the general theory of RKHS. Much of the early part of this book is an expansion of his work.

The fascination with the subject of RKHS stems from the intrinsic beauty of the field together with the remarkable number of areas in which they play a role. The theory of RKHS appears in complex analysis, group representation theory, metric embedding theory, statistics, probability, the study of integral operators, and many other areas of analysis. It is for this reason that in our book the theory is complemented by numerous and varied examples and applications.

In this book we attempt to present this beautiful theory to as wide an audience as possible. For this reason we have tried to keep much of the book as self-contained as we could. This led us to rewrite considerable parts of the theory, and experts will recognize that many proofs that appear here are novel.

Our book is composed of two parts.

In Part I we present the fundamental theory of RKHS. We have attempted to make the theory accessible to anyone with a basic knowledge of Hilbert spaces. However, many interesting examples require some background in complex variables and measure theory, so the reader might find that they can follow the theory, but then need further knowledge for the details of some examples.

In Part II we present a variety of applications of RKHS theory. We hope that these applications will be interesting to a broad group of readers, and for this reason we have again tried to make the presentation as accessible as possible. For example, our chapter on integral operators gives a proof of Mercer's theorem that assumes no prior knowledge of compact operators. Similarly, our presentation of stochastic processes assumes less probability and measure theory than is generally used. We do not go into great depth on any of these particular topics, but instead attempt to convey the essence of why the theory of RKHS plays a central role in each of these topics. We try to show the ways in which a knowledge of the RKHS theory provides a common theme that unifies our understanding of each of these areas.

Regrettably, one of the authors' favorite topics, Nevanlinna-Pick interpolation, is missing from Part II. The choice was deliberate. There is already an outstanding book on the subject by Agler and McCarthy [1] and a serious presentation of the Nevanlinna-Pick theory requires more function theory than we had previously assumed. Since RKHS theory is so broad, it is likely that the reader's favorite topic has been omitted.

This book can be used in three ways. First, it can be used as a textbook for a one-semester course in RKHS. Part I, together with some of the applications in Part II, has been the foundation of a one-semester graduate topics course at the University of Houston on several occasions and can easily be covered in that time. Second, this book can be used as a self-study guide for advanced undergraduates and beginning graduate students in mathematics, computer science and engineering. Third, the book can serve as a reference on the theory of RKHS for experts in the field.

A number of people have provided useful feedback on the book that has (we hope!) improved our exposition. In particular, we would like to thank Jennifer Good from the University of Iowa for carefully reading many parts of a nearly final draft of this book and providing detailed comments. In addition, we would like to thank our many students and colleagues at the University of Houston who sat through the earlier versions of this book and pointed out omissions and errors. Ultimately we, the authors, take responsibility for any errors or oversights that remain.

The authors would also like thank the institutions that provided support during the writing of this book, namely The University of Houston, The United States Naval Academy and Vanderbilt University.

Finally, none of this would have been possible without the patience of our wives and children. Special thanks go to our dining room table, which gave up its traditional role in life to serve as a book project center.

Part I
General theory

1

Introduction

Reproducing kernel Hilbert spaces arise in a number of areas, including approximation theory, statistics, machine learning theory, group representation theory and various areas of complex analysis. However, to understand their structure and theory, the reader only needs a background in the theory of Hilbert spaces and real analysis.

In this chapter, we introduce the reader to the formal definition of the reproducing kernel Hilbert space and present a few of their most basic properties. This beautiful theory is filled with important examples arising in many diverse areas of mathematics and, consequently, our examples often require mathematics from many different areas. Rather than keep the reader in the dark by insisting on pedagogical purity, we have chosen to present examples that require considerably more mathematics than we are otherwise assuming. So the reader should not be discouraged if they find that they do not have all the prerequisites to fully appreciate the examples. The examples are intended to give a glimpse into the many areas where these spaces arise.

1.1 Definition

We will consider Hilbert spaces over the field of either real numbers, $\mathbb{R}$, or complex numbers, $\mathbb{C}$. We will use $\mathbb{F}$ to denote either $\mathbb{R}$ or $\mathbb{C}$, so that when we wish to state a definition or result that is true for either real or complex numbers, we will use $\mathbb{F}$.

Let X be a set. We denote by $\mathcal{F}(X, \mathbb{F})$ the set of functions from X to $\mathbb{F}$. The set $\mathcal{F}(X, \mathbb{F})$ is a vector space over the field $\mathbb{F}$ with the operations of addition, $(f + g)(x) = f(x) + g(x)$, and scalar multiplication, $(\lambda \cdot f)(x) = \lambda \cdot (f(x))$.

Definition 1.1. Let X be a set. We will call a subset $\mathcal{H} \subseteq \mathcal{F}(X, \mathbb{F})$ a $\boxed{\text{REPRODUCING KERNEL HILBERT SPACE}}$ or, more briefly, an $\boxed{\text{RKHS}}$ on X if

(i) $\mathcal{H}$ is a vector subspace of $\mathcal{F}(X, \mathbb{F})$;
(ii) $\mathcal{H}$ is endowed with an inner product, $\langle , \rangle$, with respect to which $\mathcal{H}$ is a Hilbert space;
(iii) for every $x \in X$, the linear $\boxed{\text{EVALUATION FUNCTIONAL}}$, $E_x : \mathcal{H} \to \mathbb{F}$, defined by $E_x(f) = f(x)$, is bounded.

If $\mathcal{H}$ is an RKHS on X, then an application of the Riesz representation theorem shows that the linear evaluation functional is given by the inner product with a unique vector in $\mathcal{H}$. Therefore, for each $x \in X$, there exists a unique vector, $k_x \in \mathcal{H}$, such that for every $f \in \mathcal{H}$, $f(x) = E_x(f) = \langle f, k_x \rangle$.

Definition 1.2. The function k_x is called the $\boxed{\text{REPRODUCING KERNEL FOR}}$ $\boxed{\text{THE POINT } x}$. The function $K : X \times X \to \mathbb{F}$ defined by

$$K(x, y) = k_y(x)$$

is called the $\boxed{\text{REPRODUCING KERNEL FOR } \mathcal{H}}$.

Note that we have

$$K(x, y) = k_y(x) = \langle k_y, k_x \rangle$$

so that

$$K(x, y) = \langle k_y, k_x \rangle = \overline{\langle k_x, k_y \rangle} = \overline{K(y, x)}$$

in the complex case and $K(x, y) = K(y, x)$ in the real case. Also,

$$\|E_y\|^2 = \|k_y\|^2 = \langle k_y, k_y \rangle = K(y, y).$$

We now look at some examples of reproducing kernel Hilbert spaces. Our examples are drawn from function theory, differential equations and statistics.

1.2 Basic examples

1.2.1 $\mathbb{C}^n$ as an RKHS

We let $\mathbb{C}^n$ denote the vector space of complex n-tuples and for $v = (v_1, \ldots, v_n)$, $w = (w_1, \ldots, w_n)$ we let

$$\langle v, w \rangle = \sum_{i=1}^n v_i \overline{w_i}$$

denote the usual inner product. Whenever we speak of $\mathbb{C}^n$ as a Hilbert space, this is the inner product that we shall intend. Of course, if we let

$X = \{1, \ldots, n\}$, then we could also think of a complex n-tuple as a function $v : X \to \mathbb{C}$, where $v(j) = v_j$. With this identification, $\mathbb{C}^n$ becomes the vector space of all functions on X. If we let $\{e_j\}_{j=1}^n$ denote the "canonical" orthonormal basis for $\mathbb{C}^n$, i.e. let e_j be the function, $e_j(i) = \begin{cases} 1, & i = j \\ 0, & i \neq j \end{cases}$, then for every $v \in \mathbb{C}^n$ we have

$$v(j) = v_j = \langle v, e_j \rangle.$$

Thus, we see that the "canonical" basis for $\mathbb{C}^n$ is precisely the set of kernel functions for point evaluations when we regard $\mathbb{C}^n$ as a space of functions. This also explains why this basis seems so much more natural than other orthonormal bases for $\mathbb{C}^n$.

Note that the reproducing kernel for $\mathbb{C}^n$ is given by

$$K(i, j) = \langle e_j, e_i \rangle = \begin{cases} 1, & i = j \\ 0, & i \neq j \end{cases},$$

which can be thought of as the identity matrix.

More generally, given any (discrete) set X, we set

$$\ell^2(X) = \{ f : X \to \mathbb{C} : \sum_{x \in X} |f(x)|^2 < +\infty \}.$$

Given $f, g \in \ell^2(X)$, we define $\langle f, g \rangle = \sum_{x \in X} f(x)\overline{g(x)}$. With these definitions $\ell^2(X)$ becomes a Hilbert space of functions on X. If for a fixed $y \in X$, we let $e_y \in \ell^2(X)$ denote the function given by $e_y(x) = \begin{cases} 1, & x = y \\ 0, & x \neq y \end{cases}$, then it is easily seen that $\{e_y\}_{y \in X}$ is an orthonormal basis for $\ell^2(X)$ and that $\langle f, e_y \rangle = f(y)$, so that these functions are also the reproducing kernels and as before

$$K(x, y) = \langle e_y, e_x \rangle = \begin{cases} 1, & x = y \\ 0, & x \neq y \end{cases}.$$

1.2.2 A non-example

Sometimes to fix ideas it helps to look at a non-example. Suppose that we take the continuous functions on $[0,1]$, $C([0, 1])$, define the usual 2-norm on this space, i.e. $\|f\|^2 = \int_0^1 |f(t)|^2 dt$, and complete to get the Hilbert space, $L^2[0, 1]$. Given any point $x \in [0, 1]$, every function in the original space $C([0, 1])$ has a value at this point, so point evaluation is well-defined on this dense subspace. Is it possible that it extends to give a bounded linear functional on $L^2[0, 1]$? Not surprisingly, the answer is no. Keeping $0 < x < 1$ fixed, the sequence

$$f_n(t) = \begin{cases} (\frac{t}{x})^n & 0 \leq t \leq x \\ (\frac{1-t}{1-x})^n & x < t \leq 1 \end{cases}$$

is in $C([0, 1])$, satisfies $f_n(x) = 1$ for all n and $\lim_n \| f_n \|_{L^2[0,1]} = 0$. This shows that the evaluation functional, although defined on this dense subspace, is unbounded for each $0 < x < 1$ and hence has no bounded extension to all of $L^2[0, 1]$. It is easy to see how to define an analogous sequence of functions for the case $x = 0$ and $x = 1$.

Thus, there is no bounded way to extend the values of functions in $C([0, 1])$ at points to regard elements in the completion $L^2[0, 1]$ as having values at points. In particular, we cannot give $L^2[0, 1]$ the structure of an RKHS on [0,1]. One of the remarkable successes of measure theory is showing that this completion can be regarded as equivalence classes of functions, modulo sets of measure 0, but this of course makes it difficult to talk about values at all points, since elements are not actual functions.

Thus, reproducing kernel Hilbert spaces are, generally, quite different from L^2-spaces.

1.3 Examples from analysis

1.3.1 Sobolev spaces on [0,1]

These are very simple examples of the types of Hilbert spaces that arise in differential equations.

A function $f : [0, 1] \to \mathbb{R}$ is $\boxed{\text{ABSOLUTELY CONTINUOUS}}$ provided that for every $\epsilon > 0$ there exists $\delta > 0$, so that when $(x_1, y_1), \ldots, (x_n, x_n)$ are any non-overlapping intervals contained in [0, 1] with $\sum_{j=1}^n |y_j - x_j| < \delta$, then $\sum_{j=1}^n |f(y_j) - f(x_j)| < \epsilon$. An important theorem from measure theory tells us that f is absolutely continuous if and only if $f'(x)$ exists for almost all x, the derivative is integrable and, up to a constant, f is equal to the integral of its derivative. Thus, absolutely continuous functions are the functions for which the first fundamental theorem of calculus applies. Let $\mathcal{H}$ denote the set of absolutely continuous functions $f : [0, 1] \to \mathbb{R}$ such that f' is square-integrable and satisfies $f(0) = f(1) = 0$. It is not hard to see that the set $\mathcal{H}$ is a vector space of functions on [0, 1].

In order to make $\mathcal{H}$ a Hilbert space we endow $\mathcal{H}$ with the nonnegative, sesquilinear form,

$$\langle f, g \rangle = \int_0^1 f'(t)g'(t)dt.$$

Let $0 \le x \le 1$ and let $f \in \mathcal{H}$. Since f is absolutely continuous we have

$$f(x) = \int_0^x f'(t)dt = \int_0^1 f'(t)\chi_{[0,x]}(t)dt.$$

Thus, by the Cauchy-Schwartz inequality,

$$|f(x)| \le \left(\int_0^1 f'(t)^2 dt\right)^{1/2} \left(\int_0^1 \chi_{[0,x]}(t)dt\right)^{1/2} = \|f\|\sqrt{x}.$$

This last inequality shows that $\langle f, f \rangle = 0$ if and only if $f = 0$. Thus, $\langle , \rangle$ is an inner product on $\mathcal{H}$. Also, for every $x \in [0, 1]$, E_x is bounded with $\|E_x\| \le \sqrt{x}$.

All that remains to show that $\mathcal{H}$ is an RKHS is to show that it is complete in the norm induced by its inner product. If $\{f_n\}$ is a Cauchy sequence in this norm, then $\{f_n'\}$ is Cauchy in $L^2[0, 1]$ and hence there exists $g \in L^2[0, 1]$ to which this sequence converges in the L^2-sense. By the above inequality, $\{f_n\}$ must be pointwise Cauchy and hence we may define a function by setting $f(x) = \lim_n f_n(x)$. Since

$$f(x) = \lim_n f_n(x) = \lim_n \int_0^x f_n'(t)dt = \int_0^x g(t)dt,$$

it follows that f is absolutely continuous and that $f' = g$, a.e.. Note that even though g was only an equivalence class of functions, $\int_0^x g(t)dt$ was independent of the particular function chosen from the equivalence class. Hence, $f' \in L^2[0, 1]$. Finally, $f(0) = \lim_n f_n(0) = 0 = \lim_n f_n(1) = f(1)$. Thus, $\mathcal{H}$ is complete and so is an RKHS on $[0, 1]$.

We now wish to find the kernel function. We know that $f(x) = \int_0^1 f'(t)\chi_{[0,x]}(t)dt$. Thus, if we could solve the boundary-value problem,

$$g'(t) = \chi_{[0,x]}(t), g(0) = g(1) = 0,$$

then $g \in \mathcal{H}$ with $f(x) = \langle f, g \rangle$ and so $g = k_x$.

Unfortunately, this boundary-value problem has no continuous solution! Yet we know the function $k_x(t)$ exists and is continuous.

Instead, to find the kernel function we formally derive a different boundary-value problem. Then we will show that the function we obtain by this formal solution belongs to $\mathcal{H}$ and we will verify that it is the kernel function.

To find $k_x(t)$, we first apply integration by parts. We have

$$f(x) = \langle f, k_x \rangle = \int_0^1 f'(t)k_x'(t)dt$$

$$= f(t)k_x'(t)\big|_0^1 - \int_0^1 f(t)k_x''(t)dt = -\int_0^1 f(t)k_x''(t)dt.$$

If we let δ_x denote the formal Dirac-delta function, then $f(x) = \int_0^1 f(t)\delta_x(t)dt$. Thus, it appears that we need to solve the boundary-value problem,

$$-k_x''(t) = \delta_x(t), k_x(0) = k_x(1) = 0.$$

The solution to this system of equations is called the $\boxed{\text{GREEN'S FUNCTION}}$ for the differential equation. Solving formally, by integrating twice and checking the boundary conditions, we find

$$K(t, x) = k_x(t) = \begin{cases} (1 - x)t, & t \leq x \\ (1 - t)x & t \geq x \end{cases}.$$

After formally obtaining this solution, it can now be easily seen that

$$k_x'(t) = \begin{cases} (1 - x) & t < x \\ -x & t > x \end{cases},$$

so k_x is differentiable except at x (thus almost everywhere), is equal to the integral of its derivative (and so is absolutely continuous), $k_x(0) = k_x(1)$ and k_x' is square-integrable. Hence, $k_x \in \mathcal{H}$. Finally, for $f \in \mathcal{H}$,

$$\langle f, k_x \rangle = \int_0^1 f'(t)k_x'(t)dt = \int_0^x f'(t)(1 - x)dt + \int_x^1 f'(t)(-x)dt$$
$$= (f(x) - f(0))(1 - x) - x(f(1) - f(x)) = f(x)$$

Thus, $k_x(t)$ satisfies all the conditions to be the reproducing kernel for $\mathcal{H}$.

Note that $\|E_x\|^2 = \|k_x\|^2 = K(x, x) = x(1 - x)$, which considerably improves our original estimate of $\|E_x\|$. The fact that we are able to obtain the precise value of $\|E_x\|$ from the fact that we had an RKHS gives some indication of the utility of this theory. It takes a pretty sharp analyst to see that the original estimate of $|f(x)| \leq \|f\|\sqrt{x}$ can be improved, but given the theory we now know that $|f(x)| \leq \|f\|\sqrt{x(1 - x)}$ and that this is the best possible inequality.

If we change $\mathcal{H}$ slightly by considering instead the space $\mathcal{H}_1$ of absolutely continuous functions $f : [0, 1] \to \mathbb{R}$ such that $f' \in L^2[0, 1]$ and $f(0) = 0$, but retain the same inner product, then $\mathcal{H}_1$ will still be an RKHS on $[0, 1]$ and will contain $\mathcal{H}$ as a codimension one subspace. In fact, every function in $\mathcal{H}_1$ can be expressed uniquely as a function in $\mathcal{H}$ plus a scalar multiple of the function $f(x) = x$.

We leave it to the exercises to verify this claim and to compute the kernel function of $\mathcal{H}_1$. Not too surprisingly, one finds that the kernel function for $\mathcal{H}_1$

is determined by a first-order boundary-value problem, instead of the second-order problem needed when two boundary conditions were specified for $\mathcal{H}$. In later chapters we will return to this example.

1.3.2 The Paley-Wiener spaces: an example from signal processing

If we imagine that a function $f : \mathbb{R} \to \mathbb{R}$ represents the amplitude $f(t)$ of a sound wave at time t, then the *Fourier transform* of f,

$$\widehat{f}(x) = \int_{-\infty}^{\infty} f(t) e^{-2\pi i x t} dt,$$

is an attempt to decompose f into "frequencies" with the number $\widehat{f}(x)$ representing the magnitude of the contribution of $e^{2\pi i x t}$. Of course, the above integral is not defined for every function f and various different sets of hypotheses can be made, depending on the situation of interest, to make this rigorous. These hypotheses often lead one to consider various reproducing kernel Hilbert spaces. One family of such examples is the Paley-Wiener spaces.

The Paley-Wiener spaces play an important role in Fourier analysis and sampling theory, especially when one wishes to study Fourier transforms of functions that are compactly supported in either the time or frequency domain.

Suppose that we wish to consider only functions that are nonzero for a finite time, say $[-A, +A]$ for some $A > 0$. If we assume that f is square- integrable on $[-A, +A]$, then the Fourier transform integral exists and we define the *Paley-Wiener space* PW_A to be the set of functions that are Fourier transforms of L^2 functions whose support is contained in the interval $[-A, +A]$. That is,

$$PW_A = \{\widehat{f} : f \in L^2[-A, +A]\}.$$

Note that even though an element $f \in L^2[-A, +A]$ is really only an equivalence class of functions, the number $\widehat{f}(x)$ is well-defined, and independent of the function chosen from the equivalence class. Thus, even though we have seen that L^2 functions can not form a reproducing kernel Hilbert space, their Fourier transforms, i.e. the space PW_A, is a well-defined set of concrete functions on $\mathbb{R}$.

We claim that, endowed with a suitable norm, the space PW_A is a reproducing kernel Hilbert space on $\mathbb{R}$. Let $F \in PW_A$, then there exists a function $f \in L^2[-A, A]$ such that

$$F(x) = \int_{-A}^{A} f(t) e^{-2\pi i x t} dt.$$

In fact, up to almost everywhere equality, there is a unique such function. That is, the linear map

$$\widehat{} : L^2[-A, +A] \to PW_A,$$

is one-to-one. To see this we use the well-known fact that the functions $e^{2\pi i n t/A}$, $n \in \mathbb{Z}$ are an orthonormal basis for $L^2[-A, +A]$. Thus, $\widehat{f}(n/A) = F(n/A) = 0$ for every $n \in \mathbb{Z}$ implies $f = 0$, a.e.. Since $\widehat{} : L^2[-A, +A] \to PW_A$ is linear, one-to-one and onto, if we define a norm on PW_A by setting

$$\|\widehat{f}\| = \|f\|_2,$$

then PW_A will be a Hilbert space and $\widehat{}$ will define a Hilbert space isomorphism. With respect to this norm we have that for any $x \in \mathbb{R}$ and $F = \widehat{f} \in PW_A$,

$$|F(x)| = \left| \int_{-A}^{A} f(t) e^{-2\pi i x t} \, dt \right| \le \|f\|_2 \|e^{2\pi i x t}\|_2 = \sqrt{2A} \|F\|.$$

Thus, PW_A is an RKHS on $\mathbb{R}$.

To compute the kernel function for PW_A, we note that since we are identifying PW_A with the Hilbert space $L^2[-A, +A]$ via the map $\widehat{}$, we have that when $F = \widehat{f}$ and $G = \widehat{g}$, then

$$\langle F, G \rangle_{PW_A} = \langle f, g \rangle_{L^2} = \int_{-A}^{+A} f(t) \overline{g(t)} dt.$$

Now since

$$\langle F, k_y \rangle_{PW_A} = F(y) = \langle f, e^{2\pi i y t} \rangle_{L^2},$$

we obtain that

$$k_y(x) = \widehat{e^{2\pi i y t}}(x) = \int_{-A}^{+A} e^{2\pi i (y-x) t} dt.$$

Evaluating this integral yields

$$K(x, y) = \begin{cases} \dfrac{1}{\pi} \dfrac{\sin(2\pi A(x - y))}{x - y} & \text{if } x \neq y \\ 2A & \text{if } x = y \end{cases}.$$

Although this approach gives us the reproducing kernel for PW_A in an efficient manner, we have not given an explicit formula for the inner product in PW_A. To obtain the inner product one needs to apply the Fourier inversion theorem and this is more detail than we wish to go into.

It is known that the Fourier transform of an L^1 function is a continuous function on $\mathbb{R}$. Since $L^2[-A, A] \subseteq L^1[-A, A]$, and is a closed subspace of $L^2(\mathbb{R})$ we see that PW_A is a Hilbert space of continuous functions on $\mathbb{R}$. In the next chapter, we will see that this last fact follows directly from the theory of reproducing kernel Hilbert spaces.

If we allow x to be complex in the Fourier transform formula, that is, if we set

$$F(z) = \int_{-\infty}^{\infty} f(t)e^{-2\pi izt}\, dt,$$

then we obtain a space of entire functions. Similar arguments show that this set of entire functions is a reproducing kernel Hilbert space of functions on $\mathbb{C}$. This space is also often denoted by PW_A. Each entire function obtained in this manner is of *exponential type,* i.e. there is a constant C such that $|F(z)| \leq Ce^{2\pi A|z|}$.

Conversely, a remarkable theorem of Paley and Wiener [13] tells us that if $F(z)$ is an entire function of exponential type satisfying this last inequality for some constants C and A as above, then $F(z)$ is an element of PW_A.

1.4 Function theoretic examples

We now consider some examples of reproducing kernel Hilbert spaces that arise in complex analysis. The Bergman spaces appearing below are named after Stefan Bergman, who originated the theory of reproducing kernel Hilbert spaces and was the earliest researcher to obtain knowledge about spaces from their kernel functions.

1.4.1 The Hardy space of the unit disk $H^2(\mathbb{D})$

This space plays a key role in function theory, operator theory and in the theory of stochastic processes.

To construct $H^2(\mathbb{D})$, we first consider formal complex power series,

$$f \sim \sum_{n=0}^{\infty} a_n z^n$$

such that $\sum_{n=0}^{\infty} |a_n|^2 < +\infty$. Using the usual definitions for sums and scalar multiples, the set of all such power series clearly forms a vector space. Given another such power series $g \sim \sum_{n=0}^{\infty} b_n z^n$ we define the inner product,

$$\langle f, g \rangle = \sum_{n=0}^{\infty} a_n \overline{b_n}.$$

Thus, we have that $\|f\|^2 = \sum_{n=0}^{\infty} |a_n|^2$. The map $L : H^2(\mathbb{D}) \to \ell^2(\mathbb{Z}^+)$ where $\mathbb{Z}^+ = \{0, 1, 2, \ldots\} = \mathbb{N} \cup \{0\}$, defined by $L(f) = (a_0, a_1, \ldots)$ is a linear inner product preserving isomorphism. Hence we see that $H^2(\mathbb{D})$ can be identified with the Hilbert space, $\ell^2(\mathbb{Z}^+)$. Thus, we see that (ii) in the definition of an RKHS is met.

Next we show that every power series in $H^2(\mathbb{D})$, converges to define a function on the disk. To see this note that if $\lambda \in \mathbb{D}$, then

$$|E_\lambda(f)| = \left|\sum_{n=0}^{\infty} a_n \lambda^n\right| \leq \sum_{n=0}^{\infty} |a_n| \, |\lambda|^n$$

$$\leq \left(\sum_{n=0}^{\infty} |a_n|^2\right)^{1/2} \left(\sum_{n=0}^{\infty} |\lambda|^{2n}\right)^{1/2} = \|f\| \cdot \frac{1}{\sqrt{1 - |\lambda|^2}}.$$

Thus, each power series defines a function on $\mathbb{D}$. Now we want to see that if two power series define the same function on $\mathbb{D}$, then they are the same power series, i.e. that their coefficients must all be equal. To see this recall that the functions that the power series define are infinitely differentiable on their radius of convergence. By differentiating the function f n times and evaluating at 0, we obtain $(n!)a_n$. Thus, if two power series are equal as functions on $\mathbb{D}$, then they are the same power series. Also, the vector space operations on formal power series clearly agree with their vector space operations as functions on $\mathbb{D}$, and so (i) is met.

The above inequality also shows that the map E_z is bounded with $\|E_z\| \leq \dfrac{1}{\sqrt{1 - |z|^2}}$ and so $H^2(\mathbb{D})$ is an RKHS on $\mathbb{D}$.

We now compute the kernel function for $H^2(\mathbb{D})$. Let $w \in \mathbb{D}$ and let $k_w \in H^2(\mathbb{D})$ denote the kernel function at the point w. Since $k_w \in H^2(\mathbb{D})$ we can write $k_w \sim \sum_{n=0}^{\infty} b_n z^n$. Let $f \sim \sum_{n=0}^{\infty} a_n z^n$ in $H^2(\mathbb{D})$. We have

$$f(w) = \sum_{n=0}^{\infty} a_n w^n = \langle f, k_w \rangle = \sum_{n=0}^{\infty} a_n \overline{b_n}.$$

Since each of the functions $z^k \in H^2(\mathbb{D})$ we see that $b_n = \overline{w^n}$. Therefore, $k_w(z) = \sum_{n=0}^{\infty} \overline{w^n} z^n$ which is in $H^2(\mathbb{D})$.

The kernel function of the Hardy space can now be computed in closed form

$$K(z, w) = k_w(z) = \sum_{n=0}^{\infty} \overline{w^n} z^n = \frac{1}{1 - \overline{w}z}.$$

The kernel function for the Hardy space is called the $\boxed{\text{SZEGÖ KERNEL}}$ on the disk. We see that $\|E_z\| = K(z, z)^{1/2} = \dfrac{1}{\sqrt{1 - |z|^2}}$, and so the above inequality was in fact sharp.

1.4.2 Bergman spaces on complex domains

Stefan Bergman introduced the concept of reproducing kernel Hilbert spaces and used them to study various problems in complex analysis. The spaces that he introduced now bear his name. Let $G \subset \mathbb{C}$ be open and connected. We let

$$B^2(G) = \{ f : G \to \mathbb{C} \, | \, f \text{ is analytic on G and } \int \int_G |f(x+iy)|^2 dxdy < +\infty \},$$

where $dxdy$ denotes area measure. We define a sesquilinear form on $B^2(G)$ by

$$\langle f, g \rangle = \int \int_G f(x+iy)\overline{g(x+iy)}dxdy.$$

If $f \in B^2(G)$ is nonzero, then since f is analytic, f is continuous and conse-quently there will be an open set on which $|f|$ is bounded away from 0. Hence, $f \neq 0$ implies that $\langle f, f \rangle > 0$, and so $B^2(G)$ is an inner product space. Also, if $f, g \in B^2(G)$ and $f = g$ a.e., then the continuous function, $|f - g|$ can not be bounded away from 0 on any open set, and so $f(z) = g(z)$ for every $z \in G$. Thus, $B^2(G)$ can be regarded as a vector subspace of $L^2(G)$.

Theorem 1.3. *Let $G \subseteq \mathbb{C}$ be open and connected. Then $B^2(G)$ is an RKHS on G.*

Proof. If we fix $w \in G$ and choose $R > 0$ such that the closed ball of radius R centered at w, $B(w; R)^-$, is contained in G, then by Cauchy's integral formula for any $0 \le r \le R$, we have $f(w) = \frac{1}{2\pi} \int_0^{2\pi} f(w + re^{i\theta})d\theta$.
    Therefore,

$$
\begin{aligned}
f(w) &= \frac{1}{\pi R^2} \int_0^R r(2\pi f(w))dr \\
&= \frac{1}{\pi R^2} \int_0^R \left(r \int_0^{2\pi} f(w + re^{i\theta})d\theta \right) dr \\
&= \frac{1}{\pi R^2} \int \int_{B(w;R)} f(w + re^{i\theta})r dr d\theta \\
&= \frac{1}{\pi R^2} \int \int_{B(w;R)} f(x + iy) \, dx \, dy.
\end{aligned}
$$

Thus, by Cauchy-Schwartz, it follows that

$$|f(w)| \le \frac{1}{\pi R^2} \|f\| \left(\int \int_{B(w;R)} dx \, dy \right)^{1/2} = \frac{1}{R\sqrt{\pi}} \|f\|.$$

This proves that for $w \in G$ the evaluation functional is bounded. So all that remains to prove that $B^2(G)$ is an RKHS is to show that $B^2(G)$ is complete in this norm. To this end let $\{f_n\} \subseteq B^2(G)$ be a Cauchy sequence. For any $w \in G$ pick R as above and pick $0 < \delta < d(B(w; R), G^c)$, where $d(\cdot, \cdot)$ denotes the distance between the two sets. Then for any z in the closed ball of radius R centered at w we have that the closed ball of radius δ centered at z is contained in G. Hence, by the above estimate, $|f_n(z) - f_m(z)| \le \frac{1}{\delta\sqrt{\pi}} \|f_n - f_m\|$. Thus, the sequence of functions is uniformly convergent on every closed ball contained in G. If we let $f(z) = \lim_n f_n(z)$ denote the pointwise limit of this sequence, then we have that $\{f_n\}$ converges uniformly to f on each closed ball contained in G and so by Montel's theorem f is analytic.

Since $B^2(G) \subseteq L^2(G)$ and $L^2(G)$ is complete, there exists $h \in L^2(G)$ such that $\|h - f_n\|_2 \to 0$. Moreover, we may choose a subsequence $\{f_{n_k}\}$ such that $h(z) = \lim_k f_{n_k}(z)$ a.e., but this implies that $h(z) = f(z)$ a.e. and so $\|f - f_n\|_2 \to 0$. Thus, $f \in B^2(G)$ and so $B^2(G)$ is complete. $\qquad\square$

Definition 1.4. Given any open connected subset $G \subseteq \mathbb{C}$, the reproducing kernel for $B^2(G)$ is called the $\boxed{\text{BERGMAN KERNEL}}$ for G.

The result that we proved above extends to open connected subsets of $\mathbb{C}^n$ and to many complex manifolds. Knowledge of the Bergman kernel of such domains has many applications in complex analysis and the study of this kernel is still an active area of research. As an example, the Bergman kernel plays a central role in the study of the curvature of such a manifold.

Note also that the above inequality shows that $B^2(\mathbb{C}) = (0)$, since in this case R could be taken as arbitrarily large, and so $|f(w)| = 0$ for any $f \in B^2(\mathbb{C})$. Thus, the only analytic function defined on the whole complex plane that is square-integrable is the 0 function.

When $A = \text{area}(G) < +\infty$, then the constant function 1 is in $B^2(G)$ and $\|1\| = \sqrt{A}$. In this case it is natural to re-normalize so that $\|1\| = 1$. To do this we just redefine the inner product to be

$$\langle f, g \rangle = \frac{1}{A} \int \int_G f(x + iy)\overline{g(x + iy)}dxdy.$$

Often, when books refer to the Bergman space on such a domain, they mean this *normalized* Bergman space. We shall adopt this convention too. So, in particular, by the space $B^2(\mathbb{D})$, we mean the space of square-integrable analytic functions on $\mathbb{D}$, with inner-product

$$\langle f, g \rangle = \frac{1}{\pi} \int \int_{\mathbb{D}} f(x + iy)\overline{g(x + iy)}dxdy.$$

1.4.3 Two multivariable examples

Given a natural number n, by a $\boxed{\text{MULTI-INDEX}}$ we mean a point, $I = (i_1, \ldots, i_n) \in (\mathbb{Z}^+)^n$. Given $z = (z_1, \ldots, z_n) \in \mathbb{C}^n$, we set $z^I = z_1^{i_1} \cdots z_n^{i_n}$. By a $\boxed{\text{POWER SERIES IN N VARIABLES}}$, we mean a formal expression of the form $f(z) = \sum_{I \in (\mathbb{Z}^+)^n} a_I z^I$, where $a_I \in \mathbb{C}$ are called the $\boxed{\text{COEFFICIENTS}}$ of f.

We define the $\boxed{\text{N-VARIABLE HARDY SPACE, } H^2(\mathbb{D}^n)}$ to be the set of all power series, $f \sim \sum_{I \in (\mathbb{Z}^+)^n} a_I z^I$, such that $\|f\|^2 = \sum_{I \in (\mathbb{Z}^+)^n} |a_I|^2 < +\infty$.

Reasoning as for the one-variable Hardy space, one can see that for each $z = (z_1, \ldots, z_n) \in \mathbb{D}^n$ the power series converges and defines an analytic function, $f(z)$ and that $H^2(\mathbb{D}^n)$ is an RKHS on $\mathbb{D}^n$ with kernel given by

$$K(z, w) = \sum_{I \in (\mathbb{Z}^+)^n} \bar{w}^I z^I = \prod_{i=1}^n \frac{1}{1 - \bar{w}_i z_i}.$$

Similarly, we can define $\boxed{\text{MULTI-VARIABLE BERGMAN SPACES, } B^2(G)}$ for $G \subset \mathbb{C}^n$ an open connected subset by using 2n-dimensional Lebesgue measure. As in the one variable case, if the Lebesgue measure of G is finite then one often uses normalized Lebesgue measure to define the norm on $B^2(G)$, so that the constant function 1 has norm 1.

1.5 Exercises

Exercise 1.1. Let $\mathcal{H}$ be a reproducing kernel Hilbert space on X. We say that $\mathcal{H}$ *separates points* provided that given $x \neq y$ in X there exists $f \in \mathcal{H}$ such that $f(x) \neq f(y)$. Prove that setting $d(x, y) = \sup\{|f(x) - f(y)| : f \in \mathcal{H}, \|f\| \leq 1\}$ defines a metric on X if and only if $\mathcal{H}$ separates points. Give a formula for $d(x, y)$ in terms of the reproducing kernel.

Exercise 1.2. Show that if $\mathcal{H}$ is an RKHS on X and $\mathcal{H}_0 \subseteq \mathcal{H}$ is a closed subspace then $\mathcal{H}_0$ is also an RKHS on X. Prove that the reproducing kernel for $\mathcal{H}_0$ for a point y is the function $P_0(k_y)$ where k_y is the reproducing kernel function for $\mathcal{H}$ and $P_0 : \mathcal{H} \to \mathcal{H}_0$ denotes the orthogonal projection of $\mathcal{H}$ onto $\mathcal{H}_0$.

Exercise 1.3. Let $\mathcal{H}$ denote the set of functions $f : [0, 1] \to \mathbb{R}$ that are absolutely continuous, with $f' \in L^2[0, 1]$ and $f(0) = f(1) = 0$, i.e. the same set of functions as in the first Sobolev space example. Define a new inner product on $\mathcal{H}$ by $\langle f, g \rangle = \int_0^1 (f(t)g(t) + f'(t)g'(t))dt$. Prove that $\mathcal{H}$ is still a Hilbert space in this new inner product, show that the kernel function is the formal solution to $-k_y'' + k_y = \delta_y, k_y(0) = k_y(1) = 0$ and find $K(x, y)$.

Exercise 1.4. Let $\mathcal{H}_1$ denote the set of functions $f : [0, 1] \to \mathbb{R}$ that are absolutely continuous with $f' \in L^2[0, 1]$ and $f(0) = 0$. Show that $\mathcal{H}_1$ is an RKHS with respect to the inner product $\langle f, g \rangle = \int_0^1 f'(t)g'(t)\, dt$ and find the kernel function.

Exercise 1.5. Let $\mathcal{H}$ denote the set of functions $f : [0, 1] \to \mathbb{R}$ that are absolutely continuous and have the property that $f(0) = f(1)$ and $f' \in L^2[0, 1]$. Show that $\mathcal{H}$ is an RKHS on $[0, 1]$ with respect to the inner product $\langle f, g \rangle = f(0)g(0) + \int_0^1 f'(t)g'(t)dt$. Find the kernel function of $\mathcal{H}$.

Exercise 1.6. Let $\mathcal{H}_N \subseteq H^2(\mathbb{D})$ denote the subspace consisting of all functions of the form $f(z) = \sum_{n=N}^{\infty} a_n z^n$. Find the reproducing kernel for $\mathcal{H}_N$.

Exercise 1.7. Show that the Bergman kernel for the normalized Bergman space $B^2(\mathbb{D})$ is given by $K(z, w) = \dfrac{1}{(1 - z\bar{w})^2}$.

Exercise 1.8. Find the reproducing kernels for the space $B^2(\mathbb{D}^2)$ with respect to the normalized and ordinary Lebesgue measure.

Exercise 1.9. Let $U = \{x + iy : 0 < x < +\infty, 0 < y < 1\}$. Give an example of a nonzero function in $B^2(U)$.

2

Fundamental results

Let X be any set and let $\mathcal{H}$ be an RKHS on X with kernel K. In this chapter we begin with a few results that show that K completely determines the space $\mathcal{H}$. We will introduce the concept of a Parseval frame and show how, given a Parseval frame for an RKHS, the kernel can be constructed as a power series. Conversely, any series that yields the kernel in this fashion must be a Parseval frame for the RKHS. Next, we will prove Moore's theorem, which characterizes the functions that are the kernel functions of some RKHS. Such functions are often called either $\boxed{\text{POSITIVE DEFINITE}}$ or $\boxed{\text{POSITIVE SEMIDEFINITE}}$. Thus, every such function yields an RKHS by Moore's theorem, but it is often quite difficult to obtain a concrete description of the induced RKHS. We call the problem of obtaining the RKHS from the function the $\boxed{\text{RECONSTRUCTION PROBLEM}}$ and we illustrate this process in some important examples.

2.1 Hilbert space structure

Proposition 2.1. *Let $\mathcal{H}$ be an RKHS on the set X with kernel K. Then the linear span of the functions, $k_y(\cdot) = K(\cdot, y)$ is dense in $\mathcal{H}$.*

Proof. A function $f \in \mathcal{H}$ is orthogonal to the span of the functions $\{k_y : y \in X\}$ if and only if $\langle f, k_y \rangle = f(y) = 0$ for every $y \in X$, which is if and only if $f = 0$. $\qquad\square$

Lemma 2.2. *Let $\mathcal{H}$ be an RKHS on X and let $\{f_n\} \subseteq \mathcal{H}$. If $\lim_n \| f_n - f \| = 0$, then $f(x) = \lim_n f_n(x)$ for every $x \in X$.*

Proof. We have $|f_n(x) - f(x)| = |\langle f_n - f, k_x \rangle| \le \| f_n - f \| \| k_x \| \to 0$. $\quad\square$

Proposition 2.3. *Let $\mathcal{H}_i$, $i = 1, 2$, be RKHSs on X with kernels K_i, $i = 1, 2$. Let $\|\cdot\|_i$ denote the norm on the space $\mathcal{H}_i$. If $K_1(x, y) = K_2(x, y)$ for all $x, y \in X$, then $\mathcal{H}_1 = \mathcal{H}_2$ and $\|f\|_1 = \|f\|_2$ for every f.*

Proof. Let $K(x, y) = K_1(x, y) = K_2(x, y)$ and let $\mathcal{W}_i = \text{span}\{k_x \in \mathcal{H}_i : x \in X\}$, $i = 1, 2$. By the above result, $\mathcal{W}_i$ is dense in $\mathcal{H}_i$, $i = 1, 2$. Note that for any $f \in \mathcal{W}_i$, we have that $f(x) = \sum_j \alpha_j k_{x_j}(x) = \sum_{j=1} \alpha_j K(x, x_j)$, which means that the values of f are independent of whether we regard it as being in $\mathcal{W}_1$ or $\mathcal{W}_2$.

Also, for such an f,

$$\|f\|_1^2 = \sum_{i,j} \alpha_i \overline{\alpha_j} \langle k_{x_i}, k_{x_j} \rangle = \sum_{i,j} \alpha_i \overline{\alpha_j} K(x_j, x_i) = \|f\|_2^2.$$

Thus, $\|f\|_1 = \|f\|_2$ for all $f \in \mathcal{W}_1 = \mathcal{W}_2$.

Finally, if $f \in \mathcal{H}_1$, then there exists a sequence of functions, $(f_n) \subseteq \mathcal{W}_1$ with $\|f - f_n\|_1 \to 0$. Since, $\{f_n\}$ is Cauchy in $\mathcal{W}_1$ it is also Cauchy in $\mathcal{W}_2$, so there exists $g \in \mathcal{H}_2$ with $\|g - f_n\|_2 \to 0$. By the above Lemma, $f(x) = \lim_n f_n(x) = g(x)$. Thus, every $f \in \mathcal{H}_1$ is also in $\mathcal{H}_2$, and by an analogous argument, every $g \in \mathcal{H}_2$ is in $\mathcal{H}_1$. Hence, $\mathcal{H}_1 = \mathcal{H}_2$.

Finally, since $\|f\|_1 = \|f\|_2$ for every f in a dense subset, we have that the norms are equal for every f. $\qquad\square$

We now look at another consequence of the above Lemma. This result gives another means of calculating the kernel for an RKHS that is very useful.

Given vectors $\{h_s : s \in S\}$ in a normed space $\mathcal{H}$, indexed by an arbitrary set S. We say that $h = \sum_{s \in S} h_s$ provided that, for every $\epsilon > 0$, there exists a finite subset $F_0 \subseteq S$, such that for any finite set F, $F_0 \subseteq F \subseteq S$, we have that $\|h - \sum_{s \in F} h_s\| < \epsilon$. Two examples of this type of convergence are given by the two **Parseval identities.** When $\{e_s : s \in S\}$ is an orthonormal basis for a Hilbert space, $\mathcal{H}$, then for any $h \in \mathcal{H}$, we have that

$$\|h\|^2 = \sum_{s \in S} |\langle h, e_s \rangle|^2$$

and

$$h = \sum_{s \in S} \langle h, e_s \rangle e_s.$$

Note that these types of sums do not need S to be an ordered set to be defined. Perhaps the key example to keep in mind is that if we set $a_n = \frac{(-1)^n}{n}$, $n \in \mathbb{N}$ then the series, $\sum_{n=1}^{\infty} a_n$ converges, but $\sum_{n \in \mathbb{N}} a_n$ does not converge. In fact, for complex numbers, one can show that $\sum_{n \in \mathbb{N}} z_n$ converges if and only if $\sum_{n=1}^{\infty} |z_n|$ converges. Thus, this convergence is equivalent to absolute convergence in the complex case.

Theorem 2.4. *Let $\mathcal{H}$ be an RKHS on X with reproducing kernel K. If $\{e_s : s \in S\}$ is an orthonormal basis for $\mathcal{H}$, then $K(x, y) = \sum_{s \in S} \overline{e_s(y)} e_s(x)$ where this series converges pointwise.*

Proof. For any $y \in X$, we have that $\langle k_y, e_s \rangle = \overline{\langle e_s, k_y \rangle} = \overline{e_s(y)}$. Hence, $k_y = \sum_{s \in S} \overline{e_s(y)} e_s$, where these sums converge in the norm on $\mathcal{H}$.

But since they converge in the norm, they converge at every point. Hence, $K(x, y) = k_y(x) = \sum_{s \in S} \overline{e_s(y)} e_s(x)$. $\qquad\square$

Theorem 2.4 allows us to recompute some of the kernels from Chapter 1.

For a quick example of this theorem, we can easily see that in the Hardy space, the functions $e_n(z) = z^n$, $n \in \mathbb{Z}^+$ form an orthonormal basis and hence, the reproducing kernel for the Hardy space is given by

$$\sum_{n \in \mathbb{Z}^+} e_n(z) \overline{e_n(w)} = \sum_{n=0}^{\infty} (z\overline{w})^n = \frac{1}{1 - z\overline{w}}.$$

Returning to our earlier example of a Sobolev space on $[0,1]$, $\mathcal{H} = \{f : [0, 1] \to \mathbb{R} : f \text{ is absolutely continuous}, f(0) = f(1) = 0, f' \in L^2[0, 1]\}$, it is easily checked that for $n \neq 0$, the functions

$$c_n(t) = \frac{1}{\sqrt{2}\pi n}(\cos(2\pi nt) - 1), \, s_n(t) = \frac{1}{\sqrt{2}\pi n} \sin(2\pi nt)$$

form an orthonormal basis for $\mathcal{H}$.

Recall that the constant function together with the functions $\sqrt{2}\cos(2\pi n\cdot)$ and $\sqrt{2}\sin(2\pi n\cdot)$ comprise the standard Fourier basis of $L^2[0, 1]$. It follows that c_n, s_n form an orthonormal set in $\mathcal{H}$. To see that this set is complete, suppose that $\langle f, s_n \rangle = \langle f, c_n \rangle$ for all $n \geq 1$. It follows that f' is perpendicular to every non-constant element of the Fourier basis. It follows that f' is constant and hence that f is a first degree polynomial. But the boundary conditions, $f(0) = f(1) = 0$, are easily seen to imply that this polynomial is 0.

Hence, we have shown that these functions are an orthonormal basis for $\mathcal{H}$. Applying the above theorem and our earlier calculation of the reproducing kernel, we have that

$$\sum_{n\geq 1} \frac{1}{2\pi^2 n^2} \big(\sin(2\pi nx) \sin(2\pi ny)$$

$$+ \cos(2\pi nx) \cos(2\pi ny) - \cos(2\pi nx) - \cos(2\pi ny) + 1 \big)$$

$$= \begin{cases} (1-y)x, & x \leq y \\ y(1-x), & x > y \end{cases}$$

$$\sum_{n\geq 1} \frac{(e^{2\pi inx} - 1)(e^{2\pi iny} - 1)}{4\pi^2 n^2}$$

$$= \sum_{n\neq 0} \frac{\cos(2\pi n(x-y)) - \cos(2\pi nx) - \cos(2\pi ny) + 1}{2\pi^2 n^2}$$

$$= \begin{cases} (1-y)x, & x \leq y \\ y(1-x), & x > y \end{cases}.$$

Theorem 2.5. *Let $\mathcal{H}$ be an RKHS on X with reproducing kernel K, let $\mathcal{H}_0 \subseteq \mathcal{H}$ be a closed subspace and let $P_0 : \mathcal{H} \to \mathcal{H}_0$ be the orthogonal projection onto $\mathcal{H}_0$. Then $\mathcal{H}_0$ is an RKHS on X with reproducing kernel $K_0(x, y) = \langle P_0(k_y), k_x \rangle$.*

Proof. Since evaluation of a point in X defines a bounded linear functional on $\mathcal{H}$, it remains bounded when restricted to the subspace $\mathcal{H}_0$. Thus, $\mathcal{H}_0$ is an RKHS on X.

Let $f \in \mathcal{H}_0$, we have

$$f(x) = \langle f, k_x \rangle = \langle P_0(f), k_x \rangle = \langle f, P_0(k_x) \rangle.$$

Hence, $P_0(k_x)$ is the kernel function for $\mathcal{H}_0$ and we have

$$K_0(x, y) = \langle P_0(k_y), P_0(k_x) \rangle = \langle P_0(k_y), k_x \rangle.$$

$\square$

If a set of functions sums to the kernel as in Theorem 2.4, it need not be an orthonormal basis for the RKHS, but such sets do have a very nice and useful characterization. The following definition is motivated by the first Parseval identity.

Definition 2.6. Let $\mathcal{H}$ be a Hilbert space with inner product $\langle \cdot, \cdot \rangle$. A set of vectors $\{f_s : s \in S\} \subseteq \mathcal{H}$ is called a **Parseval frame** for $\mathcal{H}$ provided that

$$\|h\|^2 = \sum_{s\in S} |\langle h, f_s \rangle|^2$$

for every $h \in \mathcal{H}$.

For example, if $\{u_s : s \in S\}$ and $\{v_t : t \in T\}$ are two orthonormal bases for $\mathcal{H}$, then the sets $\{u_s : s \in S\} \cup \{0\}$ and $\{u_s/\sqrt{2} : s \in S\} \cup \{v_t/\sqrt{2} : t \in T\}$ are both Parseval frames for $\mathcal{H}$.

Thus, in particular, we see that Parseval frames do not need to be linearly independent sets.

The following result shows one of the most common ways that Parseval frames arise.

Proposition 2.7. *Let $\mathcal{H}$ be a Hilbert space, let $\mathcal{M} \subseteq \mathcal{H}$ be a closed subspace and let P denote the orthogonal projection of $\mathcal{H}$ onto $\mathcal{M}$. If $\{e_s : s \in S\}$ is an orthonormal basis for $\mathcal{H}$, then $\{P(e_s) : s \in S\}$ is a Parseval frame for $\mathcal{M}$.*

Proof. Let $h \in \mathcal{M}$, then $h = P(h)$ and hence, $\langle h, e_s \rangle = \langle P(h), e_s \rangle = \langle h, P(e_s) \rangle$. Thus, $\|h\|^2 = \sum_{s \in S} |\langle h, P(e_s) \rangle|^2$ and the result follows. $\square$

The following result shows that either of the Parseval identities could have been used to define Parseval frames. Let $\ell^2(S) = \{g : S \to \mathbb{C} : \sum_{s \in S} |g(s)|^2 < \infty\}$ denote the Hilbert space of square-summable functions on S and let

$$e_t : S \to \mathbb{C}, e_t(s) = \begin{cases} 1 & t = s \\ 0 & t \neq s \end{cases}$$

be the canonical orthonormal basis.

Proposition 2.8. *Let $\mathcal{H}$ be a Hilbert space and let $\{f_s : s \in S\} \subseteq \mathcal{H}$. Then $\{f_s : s \in S\}$ the following are equivalent*

1. *The set $\{f_s : s \in S\}$ is a Parseval frame;*
2. *The function $V : \mathcal{H} \to \ell^2(S)$ given by $(Vh)(s) = \langle h, f_s \rangle$ is a well-defined isometry;*
3. *For all $h \in \mathcal{H}$, we have $h = \sum_{s \in S} \langle h, f_s \rangle f_s$.*

Moreover, if $\{f_s : s \in S\}$ is a Parseval frame, then for any $h_1, h_2 \in \mathcal{H}$, we have that $\langle h_1, h_2 \rangle = \sum_{s \in S} \langle h_1, f_s \rangle \langle f_s, h_2 \rangle$.

Proof. First assume that $\{f_s : s \in S\}$ is a Parseval frame and define $V : \mathcal{H} \to \ell^2(S)$, by $(Vh)(s) = \langle h, f_s \rangle$, so that in terms of the basis, $Vh = \sum_{s \in S} \langle h, f_s \rangle e_s$. Since f_s is a Parseval frame we see that

$$\|Vh\|^2 = \sum_{s \in S} |\langle h, f_s \rangle|^2 = \|h\|^2.$$

Now suppose that V is an isometry. Note that $\langle h, V^*e_t \rangle = \langle Vh, e_t \rangle = \langle h, f_t \rangle$, and hence, $V^*e_t = f_t$. Since V is an isometry, we have that $V^*V = I_{\mathcal{H}}$. It now follows that

$$h = V^*Vh = V^* \left(\sum_{s \in S} \langle h, f_s \rangle e_s \right) = \sum_{s \in S} \langle h, f_s \rangle V^*(e_s) = \sum_{s \in S} \langle h, f_s \rangle f_s,$$

for every $h \in \mathcal{H}$.

Finally assume that $\sum_{s \in S} \langle h, f_s \rangle f_s = h$ for all $h \in \mathcal{H}$. We have $\langle h, h \rangle = \sum_{s \in S} \langle h, f_s \rangle \langle f_s, h \rangle = \sum_{s \in S} |\langle h, f_s \rangle|^2$.

Finally, since V is an isometry, for any $h_1, h_2 \in \mathcal{H}$, we have that

$$\langle h_1, h_2 \rangle_{\mathcal{H}} = \langle V^*Vh_1, h_2 \rangle_{\mathcal{H}} = \langle Vh_1, Vh_2 \rangle_{\ell^2(S)}$$
$$= \sum_{s \in S} (Vh_1)(s)\overline{(Vh_2)(s)} = \sum_{s \in S} \langle h_1, f_s \rangle \langle f_s, h_2 \rangle.$$

$\square$

The proof of the above result shows that our first proposition about how one could obtain a Parseval frame is really the most general example.

Proposition 2.9 (Larson). *Let $\{f_s : s \in S\}$ be a Parseval frame for a Hilbert space $\mathcal{H}$, then there is a Hilbert space $\mathcal{K}$ containing $\mathcal{H}$ as a subspace and an orthonormal basis $\{e_s : s \in S\}$ for $\mathcal{K}$, such that $f_s = P_{\mathcal{H}}(e_s), s \in S$, where $P_{\mathcal{H}}$ denotes the orthogonal projection of $\mathcal{K}$ onto $\mathcal{H}$.*

Proof. Let $\mathcal{K} = \ell^2(S)$ and let $V : \mathcal{H} \to \ell^2(S)$ be the isometry of the last proposition. Identifying $\mathcal{H}$ with $V(\mathcal{H})$ we may regard $\mathcal{H}$ as a subspace of $\ell^2(S)$. Note that $P = VV^* : \ell^2(S) \to \ell^2(S)$ satisfies $P = P^*$ and $P^2 = (VV^*)(VV^*) = V(V^*V)V^* = VV^* = P$. Thus, P is the orthogonal projection onto some subspace of $\ell^2(S)$. Since $Pe_s = V(V^*e_s) = Vf_s \in V(\mathcal{H})$, we see that P is the projection onto $V(\mathcal{H})$ and that when we identify h with Vh, we have that P is the projection onto $\mathcal{H}$ with $Pe_s = Vf_s = f_s$. $\square$

The following result gives us a more general way to compute reproducing kernels than 2.4. It was first pointed out to us by M. Papadakis.

Theorem 2.10 (Papadakis). *Let $\mathcal{H}$ be an RKHS on X with reproducing kernel K and let $\{f_s : s \in S\} \subseteq \mathcal{H}$. Then $\{f_s : s \in S\}$ is a Parseval frame for $\mathcal{H}$ if and only if $K(x, y) = \sum_{s \in S} f_s(x)\overline{f_s(y)}$, where the series converges pointwise.*

Proof. Assuming that the set is a Parseval frame we have that $K(x, y) = \langle k_y, k_x \rangle = \sum_{s \in S} \langle k_y, f_s \rangle \langle f_s, k_x \rangle = \sum_{s \in S} = \sum_{s \in S} \overline{f_s(y)} f_s(x)$.

Conversely, assume that the functions sum to give K as above. If α_j are scalars and $h = \sum_j \alpha_j k_{y_j}$ is any finite linear combination of kernel functions, then

$$
\begin{aligned}
\|h\|^2 &= \sum_{i,j} \alpha_j \overline{\alpha_i} \langle k_{y_j}, k_{y_i} \rangle = \sum_{i,j} \alpha_j \overline{\alpha_i} K(y_i, y_j) \\
&= \sum_{i,j} \alpha_j \overline{\alpha_i} \sum_{s \in S} \overline{f_s(y_j)} f_s(y_i) = \sum_{i,j} \alpha_j \overline{\alpha_i} \sum_{s \in S} \langle k_{y_j}, f_s \rangle \langle f_s, k_{y_i} \rangle \\
&= \sum_{s \in S} \langle \sum_j \alpha_j k_{y_j}, f_s \rangle \langle f_s, \sum_i \alpha_i k_{y_i} \rangle = \sum_{s \in S} |\langle h, f_s \rangle|^2.
\end{aligned}
$$

By Proposition 2.1, if we let $\mathcal{L}$ denote the linear span of the kernel functions, then $\mathcal{L}$ is dense in $\mathcal{H}$. Thus, $\tilde{V} : \mathcal{L} \to \ell^2(S)$ defined by $(\tilde{V}h)(s) = \langle h, f_s \rangle$ is an isometry on $\mathcal{L}$. Hence, $\tilde{V}$ extends to be an isometry V on $\mathcal{H}$ given by the same formula. Thus, the condition to be a Parseval frame is met by the set $\{f_s : s \in S\}$. □

Remark 2.11. Later, in Exercise 3.7, we will show that the hypothesis that each $f_s \in \mathcal{H}$ is redundant. That is, we will show that the fact that $K(x, y) = \sum_{s \in S} f_s(x) \overline{f_s(y)}$ automatically implies that $f_s \in \mathcal{H}$.

2.2 Characterization of reproducing kernels

We now turn our attention to obtaining necessary and sufficient conditions for a function $K : X \times X \to \mathbb{C}$ to be the reproducing kernel for some RKHS. We first recall some facts about matrices.

Let $A = (a_{i,j})$ be a $n \times n$ complex matrix. Then A is $\boxed{\text{POSITIVE}}$ if and only if for every $\alpha_1, \ldots, \alpha_n \in \mathbb{C}$ we have that $\sum_{i,j=1}^n \overline{\alpha_i} \alpha_j a_{i,j} \geq 0$. We denote this by $A \geq 0$.

Some remarks are in order. If we let $\langle \cdot, \cdot \rangle$ denote the usual inner product on $\mathbb{C}^n$, then in terms of the inner product, $A \geq 0$ if and only if $\langle Ax, x \rangle \geq 0$ for every $x \in \mathbb{C}^n$. In fact the sum in the definition is $\langle Ax, x \rangle$ for the vector x whose i-th component is the number α_i.

We assume at this point that the reader is familiar with some of the basic facts about positive matrices. In particular, we shall use without proof that if $A \geq 0$ and $B \geq 0$ are both $n \times n$ matrices, then $A + B \geq 0$ and $rA \geq 0$ for any $r \in \mathbb{R}^+$. We will also use that positive matrices can be characterized in terms of their eigenvalues. A matrix $A \geq 0$ if and only if $A = A^*$ and every eigenvalue of A is nonnegative. For this reason, some authors might prefer to call such matrices $\boxed{\text{POSITIVE SEMIDEFINITE}}$ or $\boxed{\text{NONNEGATIVE}}$, but we stick to the notation and terminology most common in operator theory and C*-algebras. In the case that $A = A^*$ and every eigenvalue of A is strictly positive then we will call A $\boxed{\text{STRICTLY POSITIVE}}$. We will denote this by $A > 0$. Since A is a $n \times n$ matrix, we see that $A > 0$ is equivalent to $A \geq 0$ and A invertible.

Definition 2.12. Let X be a set and let $K : X \times X \to \mathbb{C}$ be a function of two variables. Then K is called a $\boxed{\text{KERNEL FUNCTION}}$ provided that for every n and for every choice of n distinct points, $\{x_1, \ldots, x_n\} \subseteq X$, the matrix $(K(x_i, x_j)) \geq 0$. We will use the notation $K \geq 0$ to denote that the function K is a kernel function.

This terminology is by no means standard. Some authors prefer to call kernel functions $\boxed{\text{POSITIVE DEFINITE FUNCTIONS}}$, while other authors call them $\boxed{\text{POSITIVE SEMIDEFINITE FUNCTIONS}}$. To further confuse matters, authors who call kernel functions positive semidefinite functions often reserve the term positive definite function for functions such that $(K(x_i, x_j))$ is strictly positive. In some areas the term kernel function is used to mean a symmetric function of two variables, i.e. $K(x, y) = K(y, x)$. Thus, some care needs to be taken when using results from different places in the literature. We have adopted our terminology to try to avoid these problems. We shall shortly prove that a function is a kernel function if and only if there is a reproducing kernel Hilbert space for which it is the reproducing kernel, which is the main reason for our choice of terminology.

Proposition 2.13. *Let X be a set and let $\mathcal{H}$ be an RKHS on X with reproducing kernel K. Then K is a kernel function.*

Proof. Fix $\{x_1, \ldots, x_n\} \subseteq X$ and $\alpha_1, \ldots, \alpha_n \in \mathbb{C}$. Then we have that $\sum_{i,j} \overline{\alpha_i} \alpha_j K(x_i, x_j) = \langle \sum_j \alpha_j k_{x_j}, \sum_i \alpha_i k_{x_i} \rangle = \| \sum_j \alpha_j k_{x_j} \|^2 \geq 0$, and the result follows. $\qquad \square$

In general, for a reproducing kernel Hilbert space the matrix $P = (K(x_i, x_j))$ is not strictly positive. But the above calculation shows that $\langle P\alpha, \alpha \rangle = 0$ if and only if $\| \sum_j \alpha_j k_{x_j} \| = 0$. Hence, for every $f \in \mathcal{H}$ we have that $\sum_j \overline{\alpha_j} f(x_j) = \langle f, \sum_j \alpha_j k_{x_j} \rangle = 0$. Thus, in this case there is an equation of linear dependence between the values of every function in $\mathcal{H}$ at this finite set of points.

Such examples do naturally exist. Recall that in the Sobolev spaces on $[0,1]$, we were interested in spaces with boundary conditions, like $f(0) = f(1)$, in which case $k_1(t) = k_0(t)$.

On the other hand, many spaces of analytic functions, such as the Hardy or Bergman spaces, contain all polynomials. Note that there is no equation of the form, $\sum_j \beta_j p(x_j) = 0$, with β_j not all zero, that is satisfied by all polynomials. Consequently, the reproducing kernels for these spaces always define matrices that are strictly positive and invertible!

Thus, for example, recalling the Szegö kernel for the Hardy space, we see that for any choice of points $\lambda_1, \ldots, \lambda_n$ in the disk, the matrix $\left(\frac{1}{1 - \bar{\lambda}_i \lambda_j} \right)$ is invertible. For one glimpse into how powerful the theory of RKHS can be, try to show this matrix is invertible by standard linear algebraic methods.

Although the above proposition is quite elementary, it has a converse that is quite deep, and this gives us the characterization of reproducing kernel functions.

Theorem 2.14 (Moore [10]). *Let X be a set and let $K : X \times X \to \mathbb{C}$ be a function. If K is a kernel function, then there exists a reproducing kernel Hilbert space $\mathcal{H}$ of functions on X such that K is the reproducing kernel of $\mathcal{H}$.*

Proof. Let $k_y : X \to \mathbb{C}$ be the function defined by $k_y(x) = K(x, y)$. From Proposition 2.1, we know that if K is the kernel function of an RKHS then the span of these functions is dense in that RKHS. So it is natural to attempt to define a sesquilinear form on the vector space that is the span of the functions k_y, with $y \in X$.

Let $W \subseteq \mathcal{F}(X)$ be the vector space of functions spanned by the set $\{ k_y : y \in X \}$. Let $B : W \times W \to \mathbb{C}$ given by $B(\sum_j \alpha_j k_{y_j}, \sum_i \beta_i k_{y_i}) = \sum_{i,j} \alpha_j \bar{\beta}_i K(y_i, y_j)$, where α_j and β_i are scalars.

Since a given function could be expressed many different ways as a sum of the functions k_y, our first task is to see that B is well-defined. To see that B is well-defined on W, it is enough to show that if $f = \sum_j \alpha_j k_{y_j}$ is identically zero as a function on X, then $B(f, g) = B(g, f) = 0$ for every $g \in W$. Since W is spanned by the functions k_y, to prove this last equation it will be enough to show that $B(f, k_y) = B(k_y, f) = 0$. But, by the definition, $B(f, k_x) = \sum_j \alpha_j K(x, y_j) = f(x) = 0$. Similarly,

$$B(k_x, f) = \sum_j \overline{\alpha_j K(y_j, x)} = \sum_j \overline{\alpha_j K(x, y_j)} = \overline{f(x)} = 0.$$

Conversely, if $B(f, w) = 0$ for every $w \in W$, then taking $w = k_y$, we see that $f(y) = 0$. Thus, $B(f, w) = 0$ for all $w \in W$ if and only if f is identically zero as a function on X.

Thus, B is well-defined and it is now easily checked that it is sesquilinear. Moreover, for any $f \in W$ we have that $f(x) = B(f, k_x)$.

Next, since $(K(y_i, y_j))$ is positive (semidefinite), for any $f = \sum_j \alpha_j k_{y_j}$, we have that $B(f, f) = \sum_{i,j} \alpha_j \overline{\alpha_i} K(y_i, y_j) \geq 0$.

Thus, we have that B defines a semidefinite inner product on W. Hence, by the same proof as for the Cauchy-Schwarz inequality, one sees that $B(f, f) = 0$ if and only if $B(g, f) = B(f, g) = 0$ for every $g \in W$. Hence we see that $B(f, f) = 0$ if and only if $f(y) = B(f, k_y) = 0$ for every $y \in X$, i.e. f is the function that is identically 0. Therefore, B is an inner product on W.

Now given any inner product on a vector space, we may complete the space by taking equivalence classes of Cauchy sequences from W to obtain a Hilbert space, $\mathcal{H}$.

We must now show that every element of $\mathcal{H}$ can be identified uniquely with a function on X (unlike, say, the case of completing the continuous functions on [0,1] in the 2-norm to get $L^2[0, 1]$, which can not be identified with functions on [0,1]).

To this end, given $h \in \mathcal{H}$ define

$$\widehat{h}(x) = \langle h, k_x \rangle$$

and let

$$\widehat{\mathcal{H}} = \{\widehat{h} : h \in \mathcal{H}\}$$

so that $\widehat{\mathcal{H}}$ is a set of functions on X. If we let $L : \mathcal{H} \to \mathcal{F}(X; \mathbb{C})$ be defined by $L(h) = \widehat{h}$, then L is clearly linear and so $\widehat{\mathcal{H}}$ is a vector space of functions on X.

Moreover, for any function $f \in W$, we have that $\widehat{f}(x) = f(x)$.

We wish to show that the map that sends $h \to \widehat{h}$ is one-to-one. That is, $\widehat{h}(x) = 0$ for all $x \in X$ if and only if $h = 0$.

Suppose that $\widehat{h}(x) = 0$ for every x. Then $h \perp k_x$ for every $x \in X$ and so $h \perp W$. Since W is dense in $\mathcal{H}$, we have that $h = 0$ and so the map $L : \mathcal{H} \to \widehat{\mathcal{H}}$ is one-to-one and onto. Thus, if we define an inner product on $\widehat{\mathcal{H}}$ by $\langle \widehat{h}_1, \widehat{h}_2 \rangle = \langle h_1, h_2 \rangle$, then $\widehat{\mathcal{H}}$ will be a Hilbert space of functions on X.

Since

$$E_x(\widehat{h}) = \widehat{h}(x) = \langle h, k_x \rangle = \langle \widehat{h}, \widehat{k_x} \rangle,$$

we see that every point evaluation is bounded and that $\widehat{k_x} = k_x$ is the reproducing kernel for the point x. Thus, $\widehat{\mathcal{H}}$ is an RKHS on X and since $\widehat{k_y}$ is the reproducing kernel for the point y, we have that $\widehat{k_y}(x) = \langle k_y, k_x \rangle = K(x, y)$ is the reproducing kernel for $\widehat{\mathcal{H}}$. $\qquad \square$

Moore's theorem, together with Proposition 2.13, shows that there is a one-to-one correspondence between RKHS on a set and kernel functions on the set.

Definition 2.15. Given a kernel function $K : X \times X \to \mathbb{C}$, we let $\mathcal{H}(K)$ denote the unique RKHS with reproducing kernel K.

Remark 2.16. When $K : X \times X \to \mathbb{R}$ is a real-valued kernel function, then in the proof of Moore's theorem we could have defined a vector space of real-valued functions on X by considering the space of real linear combinations of the functions k_y, $y \in X$. If we denote this vector space of functions by $W_{\mathbb{R}}$ and let W denote the space of complex linear combinations that appears in the proof of Moore's theorem, then it is easily seen that $W = W_{\mathbb{R}} + i W_{\mathbb{R}}$. If we continue the proof of Moore's theorem using $W_{\mathbb{R}}$ in place of W we would obtain an RKHS of real-valued functions, which we will denote by $\mathcal{H}_{\mathbb{R}}(K)$. In this case, as sets of functions, $\mathcal{H}(K) = \mathcal{H}_{\mathbb{R}}(K) + i \mathcal{H}_{\mathbb{R}}(K)$.

Note that, although K is real-valued, the definition of kernel function requires that for all sets $\{x_1, \ldots, x_n\} \subseteq X$, and all choices of *complex* scalars we have $\sum_{i,j=1}^{n} \overline{\alpha_i} \alpha_j K(x_i, x_j) \geq 0$. For example, when $X = \{1, 2\}$ and we define $K : X \times X \to \mathbb{R}$ by $K(1, 1) = K(2, 2) = 0$, $K(1, 2) = +1$, $K(2, 1) = -1$ then

$$\sum_{i,j=1}^{2} \alpha_i \alpha_j K(i, j) = \alpha_2 \alpha_1 - \alpha_1 \alpha_2 \geq 0$$

for all choices of real scalars, but K is *not* a real-valued kernel on X.

2.3 The Reconstruction Problem

One of the more difficult problems in the theory of RKHS is to start with a kernel function K on some set X and give a concrete description of the space $\mathcal{H}(K)$. We shall refer to this as the $\boxed{\text{RECONSTRUCTION PROBLEM}}$. For example, if we started with the Szegö kernel on the disk, $K(z, w) = 1/(1 - \overline{w}z)$, then the space W that we obtain in the proof of Moore's theorem consists of linear combinations of the functions $k_w(z)$, which are rational functions with a single pole of order one outside the disk. Thus, the space W will contain no polynomials. Yet the space $\mathcal{H}(K) = H^2(\mathbb{D})$ contains the polynomials as a dense subset and has the set $\{z^n : n \geq 1\}$ as an orthonormal basis.

In this section we look at a few applications of Moore's theorem where we examine the reconstruction problem.

First, we begin with a couple of general theorems that give us information about $\mathcal{H}(K)$ from K.

Theorem 2.17. *Let X be a topological space, let $X \times X$ have the product topology and let $K : X \times X \to \mathbb{C}$ be a kernel function. If K is continuous, then every function in $\mathcal{H}(K)$ is continuous.*

Proof. Let $f \in \mathcal{H}(K)$ and fix $y_0 \in X$. Given $\epsilon > 0$, we must prove that there is a neighborhood U of y_0 such that for every $y \in U$, $|f(y) - f(y_0)| < \epsilon$. By the continuity of K, we may choose a neighborhood $V \subseteq X \times X$ of (y_0, y_0) such that $(x, y) \in V$ implies that

$$|K(x, y) - K(y_0, y_0)| < \frac{\epsilon^2}{3(\|f\|^2 + 1)}.$$

Since $X \times X$ is endowed with the product topology, we may pick a neighborhood $U \subseteq X$ of y_0 such that $U \times U \subseteq V$.

For $y \in U$, we have that

$$\begin{aligned}
\|k_y - k_{y_0}\|^2 &= K(y, y) - K(y, y_0) - K(y_0, y) + K(y_0, y_0) \\
&= [K(y, y) - K(y_0, y_0)] - [K(y, y_0) - K(y_0, y_0)] \\
&\quad - [K(y_0, y) - K(y_0, y_0)] \\
&< \frac{\epsilon^2}{\|f\|^2 + 1},
\end{aligned}$$

and hence,

$$|f(y) - f(y_0)| = |\langle f, k_y - k_{y_0} \rangle| \leq \|f\| \|k_y - k_{y_0}\| < \epsilon,$$

which completes the proof of the theorem. $\qquad\qquad\square$

In the next result, we observe that the complex conjugate of a kernel function is again a kernel function and we describe the correspondence between the two spaces.

Proposition 2.18. *Let $K : X \times X \to \mathbb{C}$ be a kernel function and let $\mathcal{H}(K)$ be the corresponding RKHS. Then the function $\overline{K}$ is also a kernel function and we have that $\mathcal{H}(\overline{K}) = \{\overline{f} : f \in \mathcal{H}(K)\}$. Moreover, the map $C : \mathcal{H}(K) \to \mathcal{H}(\overline{K})$ defined by $C(f) = \overline{f}$ is a surjective conjugate-linear isometry.*

Proof. Given $x_1, \ldots, x_n \in X$ and $\alpha_1, \ldots, \alpha_n \in \mathbb{C}$, we have that

$$\sum_{i,j=1}^{n} \overline{\alpha_i} \alpha_j \overline{K}(x_i, x_j) = \overline{\sum_{i,j=1}^{n} \alpha_i \overline{\alpha_j} K(x_i, x_j)} \geq 0,$$

since K is a kernel function. Thus, $\overline{K}$ is a kernel function.

From the proof of Moore's theorem, we see that the linear span of the functions $\overline{K}(\cdot, y) = \overline{k}_y(\cdot)$ is dense in $\mathcal{H}(\overline{K})$. First, we want to show that for any choice of points $y_1, \ldots, y_n \in X$ and scalars $\alpha_1, \ldots, \alpha_n \in \mathbb{C}$, setting

$$C(\sum_{j=1}^{n} \alpha_j k_{y_j}) = \sum_{j=1}^{n} \overline{\alpha_j} \overline{k}_{y_j}$$

yields a well-defined isometric conjugate linear map on these dense linear spans.

To show that C is well-defined it is enough to show that if we had two different ways to express the same function in $\mathcal{H}(K)$ as a linear combination of the functions k_y, then the corresponding linear combinations of the functions $\overline{k}_y$ with the conjugate linear coefficients would give the same function in $\mathcal{H}(\overline{K})$. But to show this it will be enough to show that if a linear combination of functions k_y adds up to the 0 function in $\mathcal{H}$, then the corresponding conjugate linear combination of the functions $\overline{k}_y$ adds up to the 0 function in $\mathcal{H}(\overline{K})$.

To this end note that

$$\| \sum_{j=1}^{n} \alpha_j \overline{k}_{y_j} \|^2_{\mathcal{H}(\overline{K})} = \sum_{i,j=1}^{n} \alpha_j \overline{\alpha_i} \langle \overline{k}_{y_j}, \overline{k}_{y_i} \rangle_{\mathcal{H}(\overline{K})}$$

$$= \sum_{i,j=1}^{n} \alpha_j \overline{\alpha_i} \overline{K}(y_i, y_j) = \overline{\sum_{i,j=1}^{n} \overline{\alpha_j} \alpha_i K(y_i, y_j)}$$

$$= \sum_{i,j=1}^{n} \overline{\alpha_j} \alpha_i \langle k_{y_j}, k_{y_i} \rangle_{\mathcal{H}(K)} = \| \sum_{j=1}^{n} \overline{\alpha_j} k_{y_j} \|^2_{\mathcal{H}(K)}$$

This calculation shows that the conjugate linear map C is isometric, so that if the first function adds up to 0, so does the second function. Thus, C is well-defined, isometric and conjugate linear on these dense subspaces.

Now it is a standard argument in functional analysis that, by taking limits in the domain, we may extend any bounded conjugate linear map from a dense subspace to the whole space. Moreover, since C is isometric, this extension to the whole space will be isometric. Since C is isometric and its range contains a dense subspace, C will map $\mathcal{H}(K)$ onto $\mathcal{H}(\overline{K})$. Note that on the original linear span, C takes a function to its complex conjugate. Hence, $\mathcal{H}(\overline{K}) = C(\mathcal{H}(K)) = \{\overline{f} : f \in \mathcal{H}(K)\}$. $\qquad \square$

The argument used in the above proof, that a map defined on linear combinations of kernel functions is automatically well-defined provided it is isometric, is important. This idea is used repeatedly throughout the theory of reproducing kernel Hilbert spaces.

We now turn our attention to some general cases in which we can describe the solution to the reconstruction problem.

2.3.1 The RKHS induced by a function

We start with an example of a kernel function that yields a one-dimensional RKHS.

Proposition 2.19. *Let X be a set, let f be a nonzero function on X and set $K(x, y) = f(x)\overline{f(y)}$. Then K is a kernel function, $\mathcal{H}(K)$ is the one-dimensional space spanned by f, and $\|f\| = 1$.*

Proof. To see that K is positive, we compute

$$\sum_{i,j} \alpha_j \overline{\alpha_i} K(x_i, x_j) = \left| \sum_i \overline{\alpha_i} f(x_i) \right|^2 \geq 0.$$

To find $\mathcal{H}(K)$, note that every function $k_y = \overline{f(y)}f$. Hence the space W, used in the proof of Moore's theorem, is just the one-dimensional space spanned by f. Since finite-dimensional spaces are automatically complete, $\mathcal{H}$ is just the span of f.

Finally, we compute the norm of f. Fix any point y such that $f(y) \neq 0$. Then $|f(y)|^2 \cdot \|f\|^2 = \|\overline{f(y)}f\|^2 = \|k_y\|^2 = \langle k_y, k_y \rangle = K(y, y) = |f(y)|^2$ and it follows that $\|f\| = 1$. $\square$

2.3.2 The RKHS of the Min function

We prove that the function

$$K : [0, +\infty) \times [0, +\infty) \to \mathbb{R}$$

defined by $K(x, y) = \min\{x, y\}$ is a kernel function and obtain some information about $\mathcal{H}_{\mathbb{R}}(K)$. This kernel function and the corresponding RKHS play an important role in the study of Brownian motion. We will return to this example in the chapter on integral operators. First a lemma is helpful.

Lemma 2.20. *Let J_n denote the $n \times n$ matrix with every entry equal to 1. Then $J_n \geq 0$, J_n has n as an eigenvalue of multiplicity one and every other eigenvalue of J_n is equal to 0.*

Proof. Given $v = (\alpha_1, \ldots, \alpha_n) \in \mathbb{C}^n$ we have that

$$\langle J_n v, v \rangle = \sum_{i,j=1}^{n} \alpha_j \overline{\alpha_i} = |\sum_{j=1}^{n} \alpha_j|^2 \geq 0,$$

hence, $J_n \geq 0$.

If we let v_1 denote the vector with every entry equal to 1, then $J_n v_1 = n v_1$ so one eigenvalue is n.

Note that $J_n^2 = n J_n$ so that if w is any nonzero eigenvector with eigenvalue λ then $\lambda^2 w = J_n^2 w = n J_n w = n \lambda w$. Hence, $\lambda^2 = n\lambda$ so that $\lambda \in \{0, n\}$. However, since the trace of J_n is the sum of all eigenvalues and this is equal to n, we must have that one eigenvalue is equal to n and the rest are 0. $\qquad \square$

Proposition 2.21. *Let $K : [0, +\infty) \times [0, +\infty) \to \mathbb{R}$ be defined by $K(x, y) = \min\{x, y\}$. Then K is a kernel function.*

Proof. Let $x_1, \ldots, x_n \in [0, +\infty)$ and consider the matrix $(K(x_i, x_j))$ which we must prove is positive. The proof is by induction on the number of points. Clearly $\min\{x, x\} = x \geq 0$, so the case of one point is done. If we permute the points such that $0 \leq x_1 \leq x_2 \leq \cdots \leq x_n$, then this corresponds to conjugating the original matrix by a permutation unitary, which does not affect whether or not the matrix is positive. So we may and do assume that the points are given in this ordering.

The matrix $(K(x_i, x_j))$ has the form

$$\begin{bmatrix} x_1 & x_1 & \cdots & x_1 \\ x_1 & x_2 & \cdots & x_2 \\ \vdots & \vdots & \ddots & \vdots \\ x_1 & x_2 & \cdots & x_n \end{bmatrix} = x_1 J_n + \begin{bmatrix} 0 & 0 & \cdots & 0 \\ 0 & x_2 - x_1 & \cdots & x_2 - x_1 \\ \vdots & \vdots & \ddots & \vdots \\ 0 & x_2 - x_1 & \cdots & x_n - x_1 \end{bmatrix},$$

where J_n is the matrix of all ones. Since $x_1 \geq 0$ and $J_n \geq 0$ the first matrix in this sum is positive. The lower $(n - 1) \times (n - 1)$ block of the second matrix is of the form $K(y_i, y_j)$ where $y_i = x_i - x_1 \geq 0$, for $i \geq 2$. By the induction hypothesis, this $(n - 1) \times (n - 1)$ matrix is positive. Since the sum of two positive matrices is positive the matrix $(K(x_i, x_j)) \geq 0$. $\qquad \square$

Thus, K induces an RKHS of continuous real-valued functions $\mathcal{H}_{\mathbb{R}}(K)$ on $[0, +\infty)$ and we would like to try to get some information about this space. First, it is not hard to show that K is continuous and hence, by our earlier result, every function in this space is continuous on $[0, +\infty)$.

We look at a typical function in $W_{\mathbb{R}}$. Choosing points $y_1 < \cdots < y_n$ in $[0, +\infty)$ and scalars $a_1, \ldots, a_n \in \mathbb{R}$, we see that a typical function is given by

$$\sum_{i=1}^{n} a_i k_{y_i}(x) = \begin{cases} (\sum_{i=1}^{n} a_i)x, & 0 \le x < y_1 \\ a_1 y_1 + (\sum_{i=2}^{n} a_i)x, & y_1 \le x < y_2 \\ \vdots & \vdots \\ (\sum_{i=1}^{n-1} a_i y_i) + a_n x, & y_{n-1} \le x < y_n \\ \sum_{i=1}^{n} a_i y_i, & y_n \le x \end{cases}.$$

Thus, we see that every function in the span of the kernel functions is continuous, piecewise linear, 0 at 0 and eventually constant. Conversely, it can be shown that every such function belongs to the span of the kernel functions. So $\mathcal{H}_{\mathbb{R}}(K)$ will be a space of continuous functions that is the completion of this space of "sawtooth" functions.

Giving a full characterization of the functions in this space is too long a detour for now, but we will return to this example in later chapters.

2.3.3 The RKHS induced by a positive matrix

Given a square complex matrix $P = (p_{i,j})_{i,j=1}^{n}$, if we let $X = \{1, \ldots, n\}$, then we may identify P with a function $K : X \times X \to \mathbb{C}$ by setting $K(i, j) = p_{i,j}$. With this identification, it is easy to see that K is a kernel function if and only if P is positive semidefinite. Thus, we see that every positive semidefinite $n \times n$ matrix P determines an RKHS of functions on X. Identifying functions on X with the vector space $\mathbb{C}^n$, by identifying a function $f : X \to \mathbb{C}$ with the n-tuple $(f(1), \ldots, f(n))$, we see that P determines an RKHS $\mathcal{H}(K)$ that is a subspace of $\mathbb{C}^n$. We wish to describe this space in some detail.

Note that the kernel for the point j is $k_j(i) = p_{i,j}$, which is the j-th column of the matrix P. By Moore's theorem we know that $\mathcal{H}(K)$ is the closed linear span of the kernel functions and hence (using the fact that all finite-dimensional spaces are automatically closed), $\mathcal{H}(K)$ is just the subspace of $\mathbb{C}^n$ spanned by the columns of P, which is the same as the range of the linear transformation determined by P, $\mathcal{R}(P)$.

Thus, as a space of functions $\mathcal{H}(K) = \mathcal{R}(P) \subseteq \mathbb{C}^n$, but possibly with a different inner product than the usual inner product on $\mathbb{C}^n$. To distinguish between these two inner products, and to simplify notation, we shall write $\langle \cdot, \cdot \rangle$ for the usual inner product and $\langle \cdot, \cdot \rangle_K$ for the inner product in $\mathcal{H}(K)$.

To describe this inner product, it is easiest to first consider the case when P is invertible so that $\mathcal{H}(K) = \mathbb{C}^n$ as vector spaces. Recall that every positive matrix has a unique positive square root, i.e. $P^{1/2}$ is positive semidefinite and

$(P^{1/2})^2 = P$. If we let $e_j, j = 1, \ldots, n$ denote the canonical orthonormal basis for $\mathbb{C}^n$, then we have that

$$\left\langle P^{1/2}e_j, P^{1/2}e_i \right\rangle = \langle Pe_j, e_i \rangle = p_{i,j} = k_j(i) = \langle k_j, k_i \rangle_K .$$

Now since $P^{1/2}$ is invertible the vectors $P^{1/2}e_j, j = 1, \ldots, n$ are linearly independent and span $\mathbb{C}^n$. Hence, we have a well-defined linear transformation $A : \mathbb{C}^n \to \mathcal{H}(K)$ defined by setting $A(P^{1/2}e_j) = k_j$. Letting $v = \sum_j \alpha_j P^{1/2}e_j$ and $w = \sum_i \beta_i P^{1/2}e_i$ be two vectors, we have that

$$\langle v, w \rangle = \sum_{i,j} \overline{\beta_i}\alpha_j \left\langle P^{1/2}e_j, P^{1/2}e_i \right\rangle = \sum_{i,j} \overline{\beta_i}\alpha_j \langle k_j, k_i \rangle_K = \langle Av, Aw \rangle_K .$$

Thus, A is an inner product preserving invertible transformation between the Hilbert spaces $\mathbb{C}^n$ and $\mathcal{H}(K)$. Finally, since k_j is just the j-th column of P, we have that $k_j = Pe_j$, and so $A(P^{1/2}e_j) = k_j = Pe_j$, which implies that $A = P^{1/2}$. Inverting the last formula, we see that for any two vectors $v, w \in \mathbb{C}^n = \mathcal{H}(K)$,

$$\langle v, w \rangle_K = \left\langle P^{-1/2}v, P^{-1/2}w \right\rangle = \left\langle P^{-1}v, w \right\rangle .$$

When P is not invertible, by arguing as in the proof of Moore's theorem, we still will have that the map A is a well-defined inner product preserving linear map from $\mathcal{R}(P^{1/2})$ to $\mathcal{H}(K) = \mathcal{R}(P)$. Diagonalizing P one sees that $\mathcal{R}(P^{1/2}) = \mathcal{R}(P) = \mathcal{N}(P)^\perp$, where $\mathcal{N}(P)$ denotes the null space of P and that on this subspace, A is given by multiplication by the matrix $P^{1/2}$. Finally, one finds that for $v, w \in \mathcal{N}(P)^\perp$,

$$\langle v, w \rangle_K = \left\langle P^\dagger v, w \right\rangle ,$$

where $P^\dagger : \mathcal{N}(P)^\perp \to \mathcal{N}(P)^\perp$ is the map defined by $P^\dagger v = w$ if and only if $Pw = v$.

2.3.4 The RKHS induced by the inner product

Next we show that, given a Hilbert space $\mathcal{L}$ with inner product $\langle \cdot, \cdot \rangle$, the function $K(x, y) = \langle x, y \rangle$ is a kernel function and describe the corresponding RKHS.

Definition 2.22. Let $\mathcal{H}$ be a Hilbert space and let $h_1, \ldots, h_n \in \mathcal{H}$. Then the $n \times n$ matrix

$$\left(\langle h_i, h_j \rangle \right)$$

is called the $\boxed{\text{GRAMMIAN}}$ of these vectors.

Proposition 2.23. *Let $\mathcal{H}$ be a Hilbert space and let $h_1, \ldots, h_n \in \mathcal{H}$. Then their Grammian, $G = \left(\langle h_i, h_j \rangle \right)$ is a positive semidefinite matrix. Moreover, G is a positive definite matrix if and only if $h_1, \ldots, h_n$ are linearly independent.*

Proof. Let $y = (y_1, \ldots, y_n)^t \in \mathbb{C}^n$ and compute

$$\langle Gy, y \rangle = \sum_{i,j=1}^n \langle h_i, h_j \rangle \, y_j \overline{y_i} = \| \sum_{i=1}^n \overline{y_i} h_i \|^2 \geq 0.$$

Also, $\langle Gy, y \rangle = 0$ if and only if the corresponding linear combination is the 0 vector. Hence, G is positive definite if and only if no nontrivial linear combination of $h_1, \ldots, h_n$ is 0. □

Proposition 2.24. *Let $\mathcal{L}$ be a Hilbert space with inner product $\langle \cdot, \cdot \rangle$ and let $K : \mathcal{L} \times \mathcal{L} \to \mathbb{C}$ be defined by $K(x, y) = \langle x, y \rangle$. Then K is a kernel function on $\mathcal{L}$, $\mathcal{H}(K)$ is the vector space of bounded linear functionals on $\mathcal{L}$ and the norm of a functional in $\mathcal{H}(K)$ is the same as its norm as a bounded linear functional.*

Proof. The previous proposition shows that K is a kernel function. Note that for each $y \in \mathcal{L}$, $k_y : \mathcal{L} \to \mathbb{C}$ is the bounded linear functional given by taking the inner product with y. Thus, linear combinations of kernel functions are again bounded linear functionals on $\mathcal{L}$. We need to see that every function in $\mathcal{H}(K)$ is of this form.

By the Riesz Representation theorem, each bounded linear functional $f : \mathcal{L} \to \mathbb{C}$ is uniquely determined by a vector $w \in \mathcal{L}$ so that $f = f_w$ where we set

$$f_w(v) = \langle v, w \rangle .$$

Note that, given a scalar $\lambda \in \mathbb{C}$, the linear functional $\lambda f_w = f_{\overline{\lambda} w}$, i.e. the space of bounded linear functionals is conjugate linearly isomorphic to $\mathcal{L}$ and is itself a Hilbert space in the inner product

$$\langle f_{w_1}, f_{w_2} \rangle = \langle w_2, w_1 \rangle .$$

Letting $\mathcal{H} = \{ f_w : w \in \mathcal{L} \}$ be the Hilbert space of bounded linear functionals on $\mathcal{L}$ we have, for each $x \in \mathcal{L}$, the evaluation map $E_x : \mathcal{H} \to \mathbb{C}$, given by $E_x(f_w) = f_w(x)$ satisfies $|E_x(f_w)| = |\langle x, w \rangle| \leq \|x\| \|f_w\|$, so each evaluation map is bounded on $\mathcal{H}$ and so $\mathcal{H}$ is an RKHS. For each $x \in X$ and each $f_w \in \mathcal{H}$, we have that

$$f_w(x) = \langle x, w \rangle = \langle f_w, f_x \rangle ,$$

so that the kernel function for evaluation at x is $k_x = f_x$. Hence, for $x, y \in X$, the kernel function for $\mathcal{H}$ is

$$K_{\mathcal{H}}(x, y) = k_y(x) = f_y(x) = \langle x, y \rangle.$$

Thus, $K_{\mathcal{H}} = K$ and so by the uniqueness of the RKHS determined by a kernel function the result follows. □

Note that if we define $C : \mathcal{L} \to \mathcal{H}(K)$ by $C(y) = k_y$, then $C(\lambda y) = k_{\lambda y} = \bar{\lambda} k_y$ since the inner product is conjugate linear in the second variable. We have that C is the usual conjugate linear identification between a Hilbert space and its dual.

2.4 Exercises

Exercise 2.1. Let $K_i : X \times X \to \mathbb{C}, i = 1, 2$ be kernel functions and let $p_i \geq 0, i = 1, 2$. Prove that $p_1 K_1 + p_2 K_2$ is a kernel function.

Exercise 2.2. Let $K : X \times X \to \mathbb{C}$ be a kernel function and let $g : X \to \mathbb{C}$ be a function. Prove that $K_1(x, y) = g(x) K(x, y) \overline{g(y)}$ is a kernel function.

Exercise 2.3. Let $K(s, t) = \min\{s, t\}$. Prove that $K : \mathbb{R} \times \mathbb{R} \to \mathbb{R}$ is continuous. Is K a kernel function on this domain? What about the function $K : [0, \infty) \times [0, \infty) \to \mathbb{C}$ given by $K(x, y) = \max\{x, y\}$?

Exercise 2.4. Let $y_0 = 0 < y_1 < \cdots < y_n$ be given and let $f : [0, +\infty) \to \mathbb{R}$ be a function satisfying:

- f is continuous;
- $f(0) = 0$;
- f is linear on each of the intervals $[y_i, y_{i+1}], 0 \leq i \leq n - 1$;
- f is constant for $x \geq y_n$.

Prove that f is a linear combination of the functions $k_{y_i}(x) = \min\{x, y_i\}$.

Exercise 2.5. Let $\mathcal{H}_{\mathbb{R}}(K)$ be the RKHS on $[0, +\infty)$ induced by the kernel function $K(x, y) = \min\{x, y\}$, let $0 \leq y_1 < \cdots < y_n$, let $a_1, \ldots, a_n \in \mathbb{R}$ and let $f(x) = \sum_{i=1}^{n} a_i k_{y_i}(x)$. Prove that the norm of f in $\mathcal{H}_{\mathbb{R}}(K)$ is given by the formula

$$\|f\|^2 = \left(\sum_{i=1}^{n} a_i\right)^2 y_1 + \left(\sum_{i=2}^{n} a_i\right)^2 (y_2 - y_1) + \cdots + a_n^2 (y_n - y_{n-1}).$$

Exercise 2.6. Let $p : \mathbb{R} \to \mathbb{R}$ be a piecewise continuous, nonnegative function with compact support. Define $K : \mathbb{R} \times \mathbb{R} \to \mathbb{C}$ by

$$K(x, y) = \int_{-\infty}^{+\infty} p(t)e^{i(x-y)t} dt.$$

Prove that K is a kernel function.

Exercise 2.7. Prove that

$$K(x, y) = \begin{cases} \dfrac{\sin(x - y)}{x - y} & \text{if } x \neq y \\ 1 & \text{if } x = y \end{cases}$$

is a kernel function on $X = \mathbb{R}$. [Hint: try letting p be the characteristic function of an interval.]

Exercise 2.8. Show that the following sets are Parseval frames in $\mathbb{C}^2$.

1. The set of three vectors, $\left\{ (\frac{1}{\sqrt{3}}, \frac{1}{\sqrt{2}}), (\frac{1}{\sqrt{3}}, 0), (\frac{1}{\sqrt{3}}, \frac{-1}{\sqrt{2}}) \right\}$.

2. For $N \geq 3$, the set of vectors, $\left\{ \sqrt{\frac{2}{N}} \left(\cos(\frac{2\pi j}{N}), \sin(\frac{2\pi j}{N}) \right) : j = 1, \dots, N \right\}$.

Exercise 2.9. Let $\{f_1, \dots, f_N\}$ be a Parseval frame for $\mathcal{H}$. Prove that if $dim(\mathcal{H}) = N$, then $\{f_1, \dots, f_N\}$ is an orthonormal basis for $\mathcal{H}$.

Exercise 2.10. Let $\mathcal{L}$ be a (complex) Hilbert space with inner product $\langle \cdot, \cdot \rangle$. Prove that the function $K : \mathcal{L} \times \mathcal{L} \to \mathbb{C}$ defined by $K(x, y) = \langle y, x \rangle$ is a kernel function, that $\mathcal{H}(K)$ is the vector space of bounded, conjugate linear functionals on $\mathcal{L}$ and that the map $C : \mathcal{L} \to \mathcal{H}(K)$ given by $C(y) = k_y$ is an onto linear isometry.

3

Interpolation and approximation

3.1 Interpolation in an RKHS

One of the primary applications of the theory of reproducing kernel Hilbert spaces is to problems of interpolation and approximation. It turns out that it is quite easy to give concrete formulas for interpolation and approximation in these spaces.

Definition 3.1. Let X and Y be sets, let $\{x_1, \ldots, x_n\} \subseteq X$ be a collection of distinct points, and let $\{\lambda_1, \ldots, \lambda_n\} \subseteq Y$ be a subsets. We say that a function $g : X \to Y$ $\boxed{\text{INTERPOLATES}}$ these points provided that $g(x_i) = \lambda_i$, for all $i = 1, \ldots, n$.

Let $\mathcal{H}$ be an RKHS on X with reproducing kernel K. Assume that $\{x_1, \ldots, x_n\} \subseteq X$ is a set of distinct points and that $\{\lambda_1, \ldots, \lambda_n\} \subseteq \mathbb{C}$ is a collection of possibly non-distinct numbers. We will give necessary and sufficient conditions for there to exist a function $g \in \mathcal{H}$ that interpolates these values and then we will prove that there is a unique such function of minimum norm and give a concrete formula for this function. We will then use this same technique to give a solution to the reconstruction problem.

Before proceeding, we adopt the following notation. Given a finite set $F = \{x_1, \ldots, x_n\} \subseteq X$ of distinct points, we will let $\mathcal{H}_F \subseteq \mathcal{H}$ denote the subspace spanned by the kernel functions, $\{k_{x_1}, \ldots, k_{x_n}\}$.

Note that $\dim(\mathcal{H}_F) \leq n$ and its dimension would be strictly less if and only if there is some nonzero equation of linear dependence among these functions. To understand what this means, suppose that $\sum_{j=1}^{n} \alpha_j k_{x_j} = 0$, then for every $f \in \mathcal{H}$,

$$0 = \langle f, \sum_{j=1}^{n} \alpha_j k_{x_j} \rangle = \sum_{j=1}^{n} \overline{\alpha_j} f(x_j).$$

Thus, we see that $\dim(\mathcal{H}_F) < n$ if and only if the values of every $f \in \mathcal{H}$ at the points in F satisfy some linear relation. This can also be seen to be equivalent to the fact that the linear map $T_F : \mathcal{H} \to \mathbb{C}^n$ defined by $T_F(f) = (f(x_1), \ldots, f(x_n))$ is not onto, since any vector $(\alpha_1, \ldots, \alpha_n)$ expressing an equation of linear dependence would be orthogonal to the range of T_F.

Thus, when $\dim(\mathcal{H}_F) < n$, there will exist $\lambda_1, \ldots, \lambda_n \in \mathbb{C}$ which cannot be interpolated by any $f \in \mathcal{H}$.

We've seen that it is possible for there to be such equations of linear dependence between the kernel functions, and sometimes when we construct a RKHS this can be a desirable property. This was the case for the Sobolev space example, where the boundary condition of the differential equation $f(0) = f(1) = 0$ requires $k_1 = k_0 = 0$. Changing the boundary conditions to $f(0) = f(1)$ would require $k_1 = k_0$.

We shall let P_F denote the orthogonal projection of $\mathcal{H}$ onto $\mathcal{H}_F$.

Note that $g \in \mathcal{H}_F^\perp$ if and only if $g(x_i) = \langle g, k_{x_i} \rangle = 0$, for all $i = 1, \ldots, n$. Hence, for any $h \in \mathcal{H}$, we have that

$$P_F(h)(x_i) = h(x_i), i = 1, \ldots, n.$$

Proposition 3.2. *Let $\{x_1, \ldots, x_n\}$ be a set of distinct points in X and let $\{\lambda_1, \ldots, \lambda_n\} \subseteq \mathbb{C}$. If there exists $g \in \mathcal{H}$ that interpolates these values, then $P_F(g)$ is the unique function of minimum norm that interpolates these values.*

Proof. By the remarks, if g_1 and g_2 are any two functions that interpolate these points, then $(g_1 - g_2) \in \mathcal{H}_F^\perp$. Thus, all possible solutions of the interpolation problem are of the form $g + h, h \in \mathcal{H}_F^\perp$ and $P_F(g)$ belongs to this set.

Note that for any $h \in \mathcal{H}_F^\perp$ we have $\|P_F(g)\| = \|P_F(g + h)\| \le \|g + h\|$ and so $P_F(G)$ is the unique vector of minimum norm that interpolates these values. $\qquad \square$

We now give necessary and sufficient conditions for the existence of a solution to this interpolation problem.

First, some comments on matrices and vectors are in order. If $A = (a_{i,j})$ is an $n \times n$ matrix and we wish to write a matrix vector equation $v = Aw$, then we need to regard v and w as column vectors. For typographical reasons it is easier to write row vectors, so we will write a typical column vector as $v = (v_1, \ldots, v_n)^t$, where, "t" denotes the transpose. Every such matrix defines a linear transformation from $\mathbb{C}^n$ to $\mathbb{C}^n$. We prefer to refer to the vector space $\{w \in \mathbb{C}^n : Aw = 0\}$ as the $\boxed{\text{NULLSPACE OF A}}$, instead of as the kernel, in order to avoid confusion, and we denote this set

by $\boxed{\mathcal{N}(A)}$. By the $\boxed{\text{RANGE OF A}}$ we mean the vector space, $\boxed{\mathcal{R}(A)}$:=
$\{Aw : w \in \mathbb{C}^n\}$.

Proposition 3.3. *Let X be a set, let $\mathcal{H}$ be an RKHS on X with kernel K and let $\{x_1, \ldots, x_n\} \subseteq X$ be a finite set of distinct points. If $w = (\alpha_1, \ldots, \alpha_n)^t$ is a vector in the nullspace of $(K(x_i, x_j))$, then the function, $f = \sum_j \alpha_j k_{x_j}$ is identically 0. Consequently, if $w_1 = (\alpha_1, \ldots, \alpha_n)^t$ and $w_2 = (\beta_1, \ldots, \beta_n)^t$ are two vectors satisfying $(K(x_i, x_j))w_1 = (K(x_i, x_j))w_2$, then $\sum_{j=1}^n \alpha_j k_{x_j}(y) = \sum_{j=1}^n \beta_j k_{x_j}(y)$ for every $y \in X$.*

Proof. We have that $f = 0$ if and only if $\|f\| = 0$. Now we compute $\|f\|^2 = \sum_{i,j} \overline{\alpha_i} \alpha_j \langle k_{x_j}, k_{x_i} \rangle = \sum_{i,j} \overline{\alpha_i} \alpha_j K(x_i, x_j) = \langle (K(x_i, x_j))w, w \rangle_{\mathbb{C}^n} = 0$, and the result follows.

To see this last remark, note that $w_1 - w_2$ is in the nullspace of the matrix and so the function $\sum_{j=1}^n (\alpha_j - \beta_j) k_{x_j}$ is identically 0. $\square$

Theorem 3.4. *(Interpolation in an RKHS) Let $\mathcal{H}$ be an RKHS on X with reproducing kernel K, let $F = \{x_1, \ldots, x_n\} \subseteq X$ be distinct points, and let $\{\lambda_1, \ldots, \lambda_n\} \subseteq \mathbb{C}$. Then there exists $g \in \mathcal{H}$ that interpolates these values if and only if $v = (\lambda_1, \ldots, \lambda_n)^t$ is in the range of the matrix $(K(x_i, x_j))$. Moreover, in this case if we choose $w = (\alpha_1, \ldots, \alpha_n)^t$ to be any vector whose image is v, then $h = \sum_i \alpha_i k_{x_i}$ is the unique function of minimal norm in $\mathcal{H}$ that interpolates these points. Moreover, $\|h\|^2 = \langle v, w \rangle$.*

Proof. First assume that there exists $g \in \mathcal{H}$ such that $g(x_i) = \lambda_i$, for all $i = 1, \ldots, n$. Then the solution of minimal norm is $P_F(g) = \sum_j \beta_j k_{x_j}$ for some scalars, $\beta_1, \ldots, \beta_n$. Since $\lambda_i = g(x_i) = P_F(g)(x_i) = \sum_j \beta_j k_{x_j}(x_i)$, we have that $w_1 = (\beta_1, \ldots, \beta_n)^t$ is a solution of $v = (K(x_i, x_j))w$.

Conversely, if $w = (\alpha_1, \ldots, \alpha_n)^t$ is any solution of the matrix vector equation $v = (K(x_i, x_j))w$ and we set $h = \sum_j \alpha_j k_{x_j}$, then h will be an interpolating function.

Note that $w - w_1$ is in the kernel of the matrix $(K(x_i, x_j))$ and hence by the above proposition, $P_F(g)$ and h are the same function. Hence, h is the function of minimal norm that interpolates these points. Finally, $\|h\|^2 = \sum_{i,j} \overline{\alpha_i} \alpha_j K(x_i, x_j) = \langle (K(x_i, x_j))w, w \rangle = \langle v, w \rangle$. $\square$

Corollary 3.5. *Let $\mathcal{H}$ be a RKHS on X with reproducing kernel K and let $F = \{x_1, \ldots, x_n\} \subseteq X$ be distinct. If the matrix $(K(x_i, x_j))$ is invertible, then for any $\{\lambda_1, \ldots, \lambda_n\} \subseteq \mathbb{C}$ there exists a function interpolating these values and the unique interpolating function of minimum norm is given by the formula,*

$g = \sum_j \alpha_j k_{x_j}$ *where* $w = (\alpha_1, \ldots, \alpha_n)^t$ *is given by* $w = (K(x_i, x_j))^{-1}v$,
with $v = (\lambda_1, \ldots, \lambda_n)^t$.

3.2 Strictly positive kernels

Given a set X, a kernel function $K : X \times X \to \mathbb{C}$ is called
STRICTLY POSITIVE if and only if, for every n and every set of distinct points
$\{x_1, \ldots, x_n\} \subseteq X$, the matrix

$$\big(K(x_i, x_j)\big)$$

is strictly positive definite, i.e.

$$\sum_{i,j=1}^{n} \overline{\alpha_i}\alpha_j K(x_i, x_j) > 0,$$

whenever $\{\alpha_1, \ldots, \alpha_n\} \subseteq \mathbb{C}$ are not all 0. From linear algebra we know that a
matrix is strictly positive definite if and only if it is positive (semi-definite) and
invertible. Many kernels that are encountered in practice have this property,
so we gather together here a few of the key properties of their corresponding RKHS. As we will see, this property is closely tied to interpolation and
separation of points.

Theorem 3.6. *Let X be a set and let $K : X \times X \to \mathbb{C}$ be a kernel. Then the
following are equivalent:*

1. *K is strictly positive;*
2. *for any n and any set of distinct points $\{x_1, \ldots, x_n\}$, the kernel functions
    $k_{x_1}, \ldots, k_{x_n}$ are linearly independent;*
3. *for any n, any set of distinct points $\{x_1, \ldots, x_n\}$, and any set $\{\alpha_1, \ldots, \alpha_n\} \subseteq
    \mathbb{C}$ that are not all 0, there exists $f \in \mathcal{H}(K)$ with*

    $$\alpha_1 f(x_1) + \cdots + \alpha_n f(x_n) \neq 0;$$

4. *for any n and any set of distinct points $\{x_1, \ldots, x_n\}$, there exist functions,
    $g_1, \ldots, g_n \in \mathcal{H}(K)$, satisfying*

    $$g_i(x_j) = \begin{cases} 1, & i = j \\ 0, & i \neq j \end{cases}.$$

Proof. Since $\sum_{i,j=1}^{n} \overline{\alpha_i}\alpha_j K(x_i, x_j) = \| \sum_{j=1}^{n} \alpha_j k_{x_j} \|^2$, the equivalence of
(1) and (2) follow.

Note that $\overline{\alpha_1}k_{x_1} + \cdots + \overline{\alpha_n}k_{x_n} = 0$ if and only if $\langle f, \sum_{i=1}^n \alpha_i k_{x_i} \rangle = 0$ for all $f \in \mathcal{H}$ if and only if $\overline{\alpha_1}f(x_1) + \cdots + \overline{\alpha_n}f(x_n) = 0$ for all $f \in \mathcal{H}$. Hence, (2) and (3) are equivalent. Thus, (1), (2) and (3) are equivalent.

To see that statement (4) implies (3), assume that $\alpha_i \neq 0$ and note that we can then choose $f = g_i$.

Finally, if (1) holds, then for each i applying Corollary 3.5 to the set of numbers $\alpha_j = 0$, $j \neq i$ and $\alpha_i = 1$, yields the function g_i. $\qquad\square$

A set of functions satisfying statement (4) is often called a $\boxed{\text{PARTITION OF}}$ $\boxed{\text{UNITY}}$ for the $\{x_1, \ldots, x_n\}$. Note that once one has a partition of unity for $x_1, \ldots, x_n$ then one gets $f \in \mathcal{H}(K)$ satisfying $f(x_i) = \lambda_i$ simply by setting, $f = \lambda_1 g_1 + \cdots + \lambda_n g_n$.

Definition 3.7. An RKHS $\mathcal{H}$ on X that satisfies any of the equivalent conditions of the above theorem is called $\boxed{\text{FULLY INTERPOLATING}}$.

We will now show that there is a way to compute a partition of unity when it exists. Assume that $F = \{x_1, \ldots, x_n\}$ and that $P = (K(x_i, x_j))$ is invertible as in the above corollary and write $P^{-1} = (b_{i,j}) = B$. Let e_j, $j = 1, \ldots, n$ denote the canonical basis vectors for $\mathbb{C}^n$. The columns of B are the unique vectors w_j, $j = 1, \ldots, n$ that are solutions to $e_j = Pw_j$. Thus, if we set

$$g_j = \sum_i b_{i,j} k_{x_i},$$

then $g_j(x_i) = \delta_{i,j}$, where $\delta_{i,j}$ denotes the Dirac delta function. Thus, these functions are a partition of unity for F. Moreover, when we set

$$g = \sum_j \lambda_j g_j$$

then g is the unique function in $\mathcal{H}_F$ satisfying $g(x_i) = \lambda_i$, $i = 1, \ldots n$. Hence, g is also the function of minimum norm interpolating these values. Thus, this particular partition of unity gives an easy means for producing the minimum norm interpolant for the given set.

For this reason, this particular partition of unity is well worth computing and is called the $\boxed{\text{CANONICAL PARTITION OF UNITY}}$ for F.

3.3 Best least squares approximants

As we saw in Theorem 3.4, if $\mathcal{H}$ is an RKHS on X, $F = \{x_1, \ldots, x_n\}$ is a finite set of distinct points and $\{\lambda_1, \ldots, \lambda_n\} \subseteq \mathbb{C}$, and the matrix $(K(x_i, x_j))$ is not invertible, then there might not exist any function $f \in \mathcal{H}$ with $f(x_i) = \lambda_i$, for all i. In these cases one is often interested in finding a function such that the LEAST SQUARE ERROR ,

$$J(f) = \sum_{i=1}^{n} |f(x_i) - \lambda_i|^2$$

is minimized and then, among all such functions, finding the function of minimum norm. As we shall see there is a unique such function and it is called the BEST LEAST SQUARES APPROXIMANT .

The following theorem proves the existence of the bet least squares approximant and gives a formula for obtaining this function.

Theorem 3.8. *Let $\mathcal{H}$ be an RKHS on X with kernel K, let $\{x_1, \ldots, x_n\} \subseteq X$ be a finite set of distinct points, let $v = (\lambda_1, \ldots, \lambda_n)^t \in \mathbb{C}^n$ and let $Q = \big(K(x_i, x_j)\big)$. Then there exists a vector $w = (\alpha_1, \ldots, \alpha_n)^t$ such that $(v - Qw) \in \mathcal{N}(Q)$. If we let*

$$g = \alpha_1 k_{x_1} + \cdots + \alpha_n k_{x_n},$$

then g minimizes the least square error and among all functions in $\mathcal{H}$ that minimize the least square error, g is the unique function of minimum norm.

Proof. Note that for any function $f \in \mathcal{H}$ there exists a vector $w \in \mathbb{C}^n$ such that $Qw = (f(x_1), \ldots, f(x_n))^t$. Hence we have $J(f) = \sum_{i=1}^{n} |f(x_i) - \lambda_i|^2 = \|Qw - v\|^2$. This is minimized for any vector $w = (\alpha_1, \ldots, \alpha_n)^t$ such that $Qw = P_{\mathcal{R}(Q)}(v) = v_1$, where $\mathcal{R}(Q)$ denotes the range of the matrix Q and $P_{\mathcal{R}(Q)}$ denotes the orthogonal projection onto this subspace.

Recall that if we choose a different vector $w' = (\alpha_1', \ldots, \alpha_n')$ that solves $v_1 = Qw'$ then we will have that $\sum_{i=1}^{n} \alpha_i k_{x_i} = \sum_{i=1}^{n} \alpha_i' k_{x_i}$, since $w - w' \in \mathcal{N}(Q)$. Since projecting a function f onto the span of the kernel functions $k_{x_1}, \ldots, k_{x_n}$ decreases the norm and does not change the value of f at the points $x_1, \ldots, x_n$ we see that g is the unique minimizer of J of smallest norm. $\square$

3.4 The elements of $\mathcal{H}(K)$

We will now use interpolation theory to give a general solution to the reconstruction problem for kernels. We will characterize the functions $f : X \to \mathbb{C}$ that belong to the RKHS $\mathcal{H}(K)$ determined by a kernel K.

The following theorem is best expressed in the language of DIRECTED SETS , NETS , and CONVERGENCE OF NETS , which we now introduce.

Formally, a directed set is any set with a partial order that has the property that, given any two elements of the set, there is always at least one element of the set that is greater than or equal to both. There is really only one directed set that will interest us. Given a set X, we let $\mathcal{F}_X$ denote the collection of all finite subsets of X. The set $\mathcal{F}_X$ is a directed set with respect to the partial order given by inclusion. Setting $F_1 \le F_2$ if and only if $F_1 \subseteq F_2$ defines a partial order on $\mathcal{F}_X$. Given any two finite sets, F_1, F_2, there is always a third finite set G that is larger than both. In particular, we could take $G = F_1 \cup F_2$. Thus, $\mathcal{F}_X$ is a directed set.

A *net* is a generalization of the concept of a sequence, but it is indexed by an arbitrary directed set. So, if $(\mathcal{F}, \le)$ is a directed set, then a net in a Hilbert space $\mathcal{H}$ is just a collection of vectors $\{g_F\}_{F \in \mathcal{F}} \subseteq \mathcal{H}$. Convergence of nets is defined by analogy with convergence of sequences. The net $\{g_F\}_{F \in \mathcal{F}}$ is said to *converge to* $g \in \mathcal{H}$, provided that for every $\epsilon > 0$ there is $F_0 \in \mathcal{F}$ such that whenever $F_0 \le F$, then $\|g - g_F\| < \epsilon$. These concepts are used in a fairly self-explanatory manner in the following results. If the reader is still uncomfortable with these notions after reading the proofs of the following results, a good reference for further reading on nets is [7].

Proposition 3.9. *Let* $\mathcal{H}$ *be an RKHS on the set* X, *let* $g \in \mathcal{H}$ *and for each finite set* $F \subseteq X$, *let* $g_F = P_F(g)$, *where* P_F *denotes the orthogonal projection of* $\mathcal{H}$ *onto* $\mathcal{H}_F$. *Then the net* $\{g_F\}_{F \in \mathcal{F}_X}$ *converges in norm to* g.

Proof. Let $K(x, y)$ denote the reproducing kernel for $\mathcal{H}$ and let $k_y(\cdot) = K(\cdot, y)$. Given $\epsilon > 0$, by Proposition 2.1, there exists a finite collection of points, $F_0 = \{x_1, \dots, x_n\}$ and scalars $\{\alpha_1, \dots, \alpha_n\}$, such that $\|g - \sum_i \alpha_i k_{x_i}\| < \epsilon$.

Since g_{F_0} is the closest point in $\mathcal{H}_{F_0}$ to g, we have that $\|g - g_{F_0}\| < \epsilon$. Now let F be any finite set, with $F_0 \subseteq F$. Then $\mathcal{H}_{F_0} \subseteq \mathcal{H}_F$ and since g_F is the closest point in $\mathcal{H}_F$ to g and $g_{F_0} \in \mathcal{H}_F$, we have that $\|g - g_F\| \le \|g - g_{F_0}\| < \epsilon$, for every $F_0 \subseteq F$, and the result follows. $\qquad\square$

Before proving the next result we will need a result about finite matrices. Recall that if A and B are self-adjoint matrices, then we write $A \le B$ or $B \ge A$ to mean that $B - A \ge 0$.

Proposition 3.10. *Let* $P \ge 0$ *be an* $n \times n$ *matrix, and let* $x = (x_1, \dots, x_n)^t$ *be a vector in* $\mathbb{C}^n$. *If* $xx^* = (x_i \overline{x_j}) \le cP$, *for some scalar,* $c > 0$, *then* $x \in \mathcal{R}(P)$. *Moreover, if* y *is any vector such that* $x = Py$, *then* $0 \le \langle x, y \rangle \le c$.

Proof. For any matrix A we have $\mathcal{N}(A^*) = \mathcal{R}(A)^\perp$ (Exercise 3.1). Since $P = P^*$ we have $\mathcal{N}(P) = \mathcal{R}(P)^\perp$. Thus, we may write $x = v + w$ with $v \in \mathcal{R}(P)$ and $w \in \mathcal{N}(P)$.

Now, $\langle x, w \rangle = \langle w, w \rangle$, and hence, $\|w\|^4 = \langle w, x \rangle \langle x, w \rangle = \sum_{i,j} \overline{x_j} w_j x_i \overline{w_i} = \langle (x_i \overline{x_j}) w, w \rangle \leq \langle cPw, w \rangle = 0$, since $Pw = 0$. This inequality shows that $w = 0$ and hence, $x = v \in \mathcal{R}(P)$.

Now if we write $x = Py$, then $\langle x, y \rangle = \langle Py, y \rangle \geq 0$. As above, we have that $\langle x, y \rangle^2 = \langle y, x \rangle \langle x, y \rangle = \langle (x_i \overline{x_j}) y, y \rangle \leq \langle cPy, y \rangle = c \langle x, y \rangle$. Cancelling one factor of $\langle x, y \rangle$ from this inequality yields the result. □

We are now able to prove a theorem that characterizes the functions that belong to an RKHS in terms of the reproducing kernel.

Theorem 3.11. *Let $\mathcal{H}$ be an RKHS on X with reproducing kernel K and let $f : X \to \mathbb{C}$ be a function. Then the following are equivalent:*

(1) $f \in \mathcal{H}$;
(2) there exists a constant, $c \geq 0$, such that for every finite subset, $F = \{x_1, \ldots x_n\} \subseteq X$, there exists a function $h \in \mathcal{H}$ with $\|h\| \leq c$ and $f(x_i) = h(x_i), i = 1, \ldots n$;
(3) there exists a constant, $c \geq 0$, such that the function, $c^2 K(x, y) - f(x)\overline{f(y)}$ is a kernel function.

Moreover, if $f \in \mathcal{H}$ then $\|f\|$ is the least c that satisfies the inequalities in (2) and (3).

Proof. *(1) implies (3):* Let $F = \{x_1, \ldots, x_n\} \subseteq X$, let $\alpha_1, \ldots, \alpha_n$ be scalars and set $g = \sum_j \alpha_j k_{x_j}$. Then $\sum_{i,j} \overline{\alpha_i} \alpha_j f(x_i) \overline{f(x_j)} = |\sum_i \overline{\alpha_i} f(x_i)|^2 = |\langle f, g \rangle|^2 \leq \|f\|^2 \|g\|^2 = \|f\|^2 \sum_{i,j} \overline{\alpha_i} \alpha_j K(x_i, x_j)$. Since the choice of the scalars was arbitrary, we have that $(f(x_i)\overline{f(x_j)}) \leq \|f\|^2 (K(x_i, x_j))$ and so (3) follows with $c = \|f\|$.

(3) implies (2): Let $F = \{x_1, \ldots, x_n\} \subseteq X$ be a finite set. Apply Proposition 3.10 to deduce that the vector v whose entries are $\lambda_i = f(x_i)$ is in the range of $(K(x_i, x_j))$. Then use the Interpolation Theorem to deduce that there exists $h = \sum_i \alpha_i k_{x_i}$ in $\mathcal{H}_F$ with $h(x_i) = f(x_i)$. Let w denote the vector whose components are the α_is and it follows that $\|h\|^2 = \langle v, w \rangle \leq c^2$ by applying Proposition 3.10 again.

(2) implies (1): By assumption, for every finite set F there exists $h_F \in \mathcal{H}$ such that $\|h_F\| \leq c$ and $h_F(x) = f(x)$ for every $x \in F$. Set $g_F = P_F(h_F)$, then $g_F(x) = h_F(x) = f(x)$ for every $x \in F$ and $\|g_F\| \leq \|h_F\| \leq c$.

We claim that the net $\{g_F\}_{F \in \mathcal{F}_X}$ is Cauchy and converges to f.

To see that the net is Cauchy, let $M = \sup \|g_F\| \leq c$ and fix $\epsilon > 0$. Choose a set F_0 such that $M - \epsilon^2 < \|g_{F_0}\|$. For any $F \in \mathcal{F}_X$ with $F_0 \subseteq F$ we have that $P_{F_0}(g_F) = g_{F_0}$ and hence, $\langle (g_F - g_{F_0}), g_{F_0} \rangle = 0$. Hence, $\|g_F\|^2 = \|g_{F_0}\|^2 + \|g_F - g_{F_0}\|^2$, and so $M - \epsilon^2 \leq \|g_{F_0}\| \leq \|g_F\| \leq M$.

Therefore, $0 \leq \|g_F\| - \|g_{F_0}\| \leq \epsilon^2$, and we have that $\|g_F - g_{F_0}\|^2 = \|g_F\|^2 - \|g_{F_0}\|^2 = (\|g_F\| + \|g_{F_0}\|)(\|g_F\| - \|g_{F_0}\|) \leq 2M\epsilon^2$. Thus, $\|g_F - g_{F_0}\| < \sqrt{2M}\epsilon$ and so for any $F_1, F_2 \in \mathcal{F}_X$ with $F_0 \subseteq F_1$, $F_0 \subseteq F_2$, it follows that $\|g_{F_1} - g_{F_2}\| < 2\sqrt{2M}\epsilon$ and we have proven that the net is Cauchy.

Thus, there is a function $g \in \mathcal{H}$ that is the limit of this net and hence, $\|g\| \leq M \leq c$. But since any norm convergent net also converges pointwise, we have that $g(x) = f(x)$ for any x. Thus, the proof that (2) implies (1) is complete.

Finally, given that $f \in \mathcal{H}$, we have that the conditions of (2) and (3) are met for $c = \|f\|$. So the least c that meets these conditions is less than $\|f\|$. Conversely, in the proof that (3) implies (2), we saw that any c that satisfies (3) satisfies (2). But in the proof that (2) implies (1), we saw that $\|f\| \leq c$. Hence any c that meets the inequalities in (2) or (3) must be greater than $\|f\|$. $\qquad\square$

The following result illustrates some of the surprising consequences of the above theorem.

Corollary 3.12. *Let $f : \mathbb{D} \to \mathbb{C}$ be a function. Then f is analytic on $\mathbb{D}$ and has a square summable power series if and only if there exists $c > 0$ such that $K(z, w) = \frac{c^2}{1 - z\overline{w}} - f(z)\overline{f(w)}$ is a kernel function on $\mathbb{D}$.*

What is a bit surprising in this last result is that the analyticity of f follows from the kernel function condition, which is just the requirement that certain matrices be positive semidefinite.

3.5 Exercises

Exercise 3.1. Let A be a $n \times k$ matrix. Prove that $\mathcal{N}(A^*) = \mathcal{R}(A)^{\perp}$.

Exercise 3.2. A reproducing kernel Hilbert space $\mathcal{H}$ on X is said to $\boxed{\text{SEPARATE POINTS}}$ provided that for every $x_1 \neq x_2$ there exists $f \in \mathcal{H}$ with $f(x_1) \neq f(x_2)$. Prove that $\mathcal{H}(K)$ separates points if and only if for every pair of distinct points $\{x_1, x_2\}$ the vector $(1, -1)^t$ is not in the nullspace of the 2×2 matrix $(K(x_i, x_j))$.

Exercise 3.3. Let $K : X \times X \to \mathbb{C}$ be a kernel function and let $x_1, \ldots, x_n$ be a set of n distinct points in X. Prove that the dimension of the span of $\{k_{x_1}, \ldots, k_{x_n}\}$ is equal to the rank of the $n \times n$ matrix $(K(x_i, x_j))$.

Exercise 3.4. Let X be a set, $K : X \times X \to \mathbb{C}$ a kernel function, and $F = \{x_1, \ldots, x_n\}$ a finite set of distinct points with $(K(x_i, x_j))$ invertible. Assume that the constant function 1 belongs to $\text{span}\{k_{x_1}, \ldots, k_{x_n}\}$. Prove that the canonical partition of unity satisfies $\sum_j g_j(x) = 1$ for all $x \in X$.

Exercise 3.5. Let K denote the Szegö kernel and let $z_1 \neq z_2$ be points in the unit disk. Compute explicitly the canonical partition of unity g_1, g_2 for $F = \{z_1, z_2\}$. What happens to these functions as $z_1 \to z_2$?

Exercise 3.6. Repeat the above problem with the Szegö kernel on the disk replaced by the Bergman kernel on the disk.

Exercise 3.7. Prove the following strengthening of Theorem 2.10. Let X and S be sets and let $f_s : X \to \mathbb{C}$, $s \in S$ be a collection of functions on X and assume that the series $K(x, y) = \sum_{s \in S} f_s(x)\overline{f_s(y)}$ converges pointwise. Then K is a kernel function, $f_s \in \mathcal{H}(K)$ for every $s \in S$ and $\{f_s : s \in S\}$ is a Parseval frame for $\mathcal{H}(K)$.

Exercise 3.8. Give the reproducing kernel condition for a function $f : [0, 1] \to \mathbb{R}$ to be absolutely continuous, with $f(0) = f(1)$ and f' square-integrable. Can you give a direct proof, without using the above theorem, from this condition that f is absolutely continuous?

Exercise 3.9. Let $x \neq y$ be points in $\mathbb{D}$. Prove that $\sup\{|f(y)| : f \in H^2(\mathbb{D}), f(x) = 0, \|f\| \leq 1\} \leq \dfrac{|x-y|}{|1-y\bar{x}|\sqrt{1-|y|^2}}$. Is this inequality sharp?

4

Cholesky and Schur

In this chapter we present some results about positive matrices that will be indispensable in understanding RKHS, but are often not included in a first course on Hilbert spaces or matrix theory. The first of these results is the basis for the Cholesky algorithm, which is, perhaps, the most efficient way to determine if a matrix is positive semidefinite and is used to obtain the Cholesky factorization of positive semidefinite matrices. This result is also the basis for the notion of Schur complements. The second result is about the Schur product (or Hadamard product, or pointwise product) of matrices. One basic fact about Schur products that we will prove in this chapter, and use often later, is that the Schur product of positive matrices is positive.

4.1 Cholesky factorization

If $A \in M_n$ is a matrix, then the matrix $P = AA^*$ is positive. The Cholesky factorization provides a remarkable converse. Not only is every positive matrix of this form, but there is in fact a lower triangular matrix L such that $P = LL^*$.

This is Cholesky's theorem. The proof is constructive, and a careful examination of it reveals an algorithm to compute L. Before deriving the Cholesky factorization we need to prove a short lemma.

Lemma 4.1. *Let $P = (p_{i,j})$ be an $n \times n$ positive semidefinite matrix. If $p_{k,k} = 0$, for some $1 \leq k \leq n$, then $p_{i,k} = p_{k,i} = 0$ for all $1 \leq i \leq n$.*

Proof. Let $x = \alpha e_k + \beta e_i$, where e_i is the ith vector in the standard basis for $\mathbb{C}^n$. We have $\langle Px, x \rangle \geq 0$, which gives

$$0 \leq |\alpha|^2 \, p_{k,k} + \beta \overline{\alpha} p_{k,i} + \overline{\beta} \alpha p_{i,k} + |\beta|^2 \, p_{i,i}.$$

This gives, $|\beta|^2 p_{i,i} \geq -\bar{\beta}\alpha p_{i,k} - \bar{\alpha}\beta p_{k,i}$ for all α, β. Now choose β such that $|\beta| = 1$ and $\beta p_{k,i} = -|p_{k,i}|$. Note that $\bar{\beta} p_{i,k} = \overline{\beta p_{k,i}} = -|p_{k,i}|$. If we restrict α to be real and use the above inequality we get $p_{i,i} \geq 2\alpha |p_{k,i}|$ for all $\alpha \in \mathbb{R}$. This can not be true for all $\alpha \in \mathbb{R}$ unless $p_{i,k} = 0$. Hence, $p_{k,i} = p_{i,k} = 0$ for $1 \leq k \leq n$. $\square$

We can now state Cholesky's useful observation.

Theorem 4.2 (Cholesky). *Let $P = (p_{i,j})$ be an $n \times n$ positive matrix and assume that $p_{k,k} \neq 0$. Then the matrix*

$$\left(p_{i,j} - \frac{p_{i,k} p_{k,j}}{p_{k,k}} \right)^n_{i,j=1}$$

is positive.

Proof. Let v be the k-th column of P, i.e. $v_i = p_{i,k}$. Note that the above matrix is $P - p_{k,k}^{-1} vv^*$. Note also that the k-th row and k-th column of this matrix is 0.

Since $p_{k,k} > 0$, it is enough to prove that the matrix $Q = (p_{k,k} p_{i,j} - p_{i,k} p_{k,j})$ is positive, i.e. for $\alpha_1, \ldots, \alpha_n \in \mathbb{C}$, we need to show that

$$\sum_{i,j=1}^n \overline{\alpha_i} \alpha_j (p_{k,k} p_{i,j} - p_{i,k} p_{k,j}) \geq 0.$$

Since P is a positive matrix we know that $\langle Px, x \rangle \geq 0$ for all $x \in \mathbb{C}^n$. By making an appropriate choice of vector x we will show that Q is positive. We set $x_i = p_{k,k}^{1/2} \alpha_i$ for $i \neq k$ and $x_k = -p_{k,k}^{-1/2} \sum_{j \neq k} p_{k,j} \alpha_j$. Since $P \geq 0$ we have that $p_{k,j} = \overline{p_{j,k}}$. We now compute $\langle Px, x \rangle$:

$$\langle Px, x \rangle = \sum_{i,j}^n \overline{x_i} x_j p_{i,j}$$

$$= \sum_{i,j \neq k}^n p_{i,j} \overline{x_i} x_j + \sum_{i \neq k}^n p_{i,k} \overline{x_i} x_k + \sum_{j \neq k}^n p_{k,j} \overline{x_k} x_j + p_{k,k} |x_k|^2$$

$$= \sum_{i,j \neq k}^n p_{k,k} p_{i,j} \overline{\alpha_i} \alpha_j + x_k \left(\sum_{i \neq k}^n p_{k,k}^{1/2} p_{i,k} \overline{\alpha_i} \right)$$

$$+ \overline{x_k} \left(\sum_{j \neq k}^n p_{k,k}^{1/2} p_{k,j} \alpha_j \right) + p_{k,k} |x_k|^2$$

$$= \sum_{i,j \neq k}^{n} p_{k,k} p_{i,j} \overline{\alpha_i} \alpha_j - p_{k,k} |x_k|^2 - p_{k,k} |x_k|^2 + p_{k,k} |x_k|^2$$

$$= \sum_{i,j \neq k}^{n} (p_{k,k} p_{i,j} - p_{i,k} p_{k,j}) \overline{\alpha_i} \alpha_j = \sum_{i,j=1}^{n} \overline{\alpha_i} \alpha_j (p_{k,k} p_{i,j} - p_{i,k} p_{k,j}),$$

where the last equality follows since the k-th row and column of Q is 0. $\square$

Whenever an $n \times n$ matrix Q is positive, then the $(n-1) \times (n-1)$ matrix Q' that one obtains by eliminating the k-th row and column, is also positive. This can most easily be seen by realizing that for any $v = (v_1, \ldots, v_{n-1}) \in \mathbb{C}^{n-1}$ we have that $\langle Q'v, v \rangle = \langle Qw, w \rangle$ where $w \in \mathbb{C}^n$ is the vector $w = (v_1, \ldots, v_{k-1}, 0, v_k, \ldots, v_{n-1})$.

Definition 4.3. The $(n-1) \times (n-1)$ positive semidefinite matrix

$$\left(p_{i,j} - \frac{p_{i,k} p_{k,j}}{p_{k,k}} \right)^n_{i,j \neq k}$$

is called the $\boxed{\text{SCHUR COMPLEMENT OF } P}$ with respect to the subspace spanned by the k-th basis vector.

It is also possible to define an analogous Schur complement of a positive semidefinite operator on a Hilbert space with respect to any subspace.

We now derive the Cholesky factorization for a positive matrix. A matrix $L = (l_{i,j})$ is called lower triangular if $l_{i,j} = 0$ for $j > i$.

Theorem 4.4 (Cholesky's Factorization). *Let P be an $n \times n$ positive semidefinite matrix. Then there exists a lower triangular matrix L such that $P = LL^*$.*

Proof. The proof is by induction. If P is a 1×1 matrix $P = [p_{1,1}]$, then we can write $P = [\sqrt{p_{1,1}}]^2$.

Now assume the result is true for $(n-1) \times (n-1)$ positive matrices. Let $P = [p_{i,j}] \in M_n$. First consider the case where $p_{1,1} = 0$. It follows that all the entries in the first row and column of P are zero, by Lemma 4.1. Now, by the induction hypothesis we can factor the matrix $Q = [p_{i,j}]^n_{i,j=2} \in M_{n-1}$ as $Q = L_1 L_1^*$. If we let $L = (l_{i,j})$ be the $n \times n$ matrix obtained from L_1 by setting $l_{i,1} = l_{1,j} = 0$ for $1 \leq i \leq n$, and $[l_{i,j}]^n_{i,j=2} = L_1$, then L is lower triangular and $P = LL^*$.

Having dealt with the case where $p_{1,1} = 0$, we may now assume that $p_{1,1} > 0$. Let v be the vector in the first column of P, i.e. $v_i = p_{i,1}$. By Cholesky's Theorem, $P - p_{1,1}^{-1} vv^*$ is a positive matrix. Note that the first row and first column of this matrix are 0. The remaining entries form a $(n - 1) \times (n - 1)$ positive semidefinite matrix Q.

Using the induction hypothesis we see that there is a $(n-1) \times (n-1)$ lower-triangular matrix $L_1 = (l_{i,j})_{i,j=2}^n$ such that $Q = L_1 L_1^*$. Let $L = [l_{i,j}]$ be the matrix given by $l_{i,1} = p_{1,1}^{-1/2} v_i$ for $1 \le i \le n$, $l_{1,j} = 0$ for $2 \le j \le n$ and $[l_{i,j}]_{i,j=2}^n = L_1$. We have $P = LL^*$.   □

4.1.1 Cholesky's algorithm

Note that the construction of the factorization $P = LL^*$ is algorithmic.

Cholesky's theorem and the resulting factorization algorithm also provide one of the fastest algorithms for determining if a matrix is positive semidefinite.

Starting with an $n \times n$ self-adjoint matrix $P = (p_{i,j})$, if $p_{1,1} < 0$ or $p_{1,1} = 0$ but some $p_{1,j} \ne 0$ then P is not positive semidefinite and we are done. When neither of these conditions happens, let's say that P *passes the basic criterion*. Now if P passes the basic criterion, we can apply the formula in Cholesky's theorem to obtain an $(n - 1) \times (n - 1)$ matrix. If this new matrix fails the basic criterion, then we know that the new matrix, and consequently P, are not positive semidefinite.

Otherwise the $(n - 1) \times (n - 1)$ matrix passes the basic criterion and we may continue inductively to obtain an $(n - 2) \times (n - 2)$ matrix.

If at any stage in this inductive process a new matrix fails the basic criterion, then P is not positive semidefinite.

Otherwise the process terminates after n steps and if we have saved the vectors from each step, then these give the entries of the lower triangular matrix L satisfying $P = LL^*$.

This process is known as $\boxed{\text{CHOLESKY'S ALGORITHM}}$. Note that it simultaneously determines if a matrix is positive semidefinite and yields the Cholesky factorization at the same time!

Also, since the size of the matrix is reduced by one at each step, the operation count is on the order of $n^2 + (n - 1)^2 + \cdots + 1^2$ which is $O(n^3)$.

Many numerical packages rely on Cholesky's algorithm, often with a simple pivoting rule, such as always choosing the largest diagonal entry, instead of just the first, for added stability.

The proof we have given of Cholesky's theorem is matrix theoretic. We can also deduce Cholesky's theorem from the theory of reproducing kernel Hilbert spaces, and we show how to do this below. The key step in the proof of the theorem is to check that the matrix Q is positive. Since P is positive, it is a kernel function on the space $X = \{1, \ldots, n\}$. The associated RKHS is a space of functions on X, i.e. vectors. The key idea is to recognize Q

as the kernel of a subspace of $\mathcal{H}(P)$. The argument applies to an arbitrary RKHS.

Theorem 4.5. *Suppose that K is a kernel on a set X. Let $x_0 \in X$ and suppose that $K(x_0, x_0) \neq 0$. Define $K_0(x, y) = K(x, y) - \dfrac{K(x, x_0) K(x_0, y)}{K(x_0, x_0)}$. Then K_0 is a kernel function, $\mathcal{H}(K_0) = \{f \in \mathcal{H}(K) : f(x_0) = 0\}$, and for $f \in \mathcal{H}(K_0)$, we have that $\|f\|_{\mathcal{H}(K_0)} = \|f\|_{\mathcal{H}(K)}$.*

A few comments before the proof. The Cholesky theorem can be used to show that K_0 is a kernel, but in the proof below we will see that it is a kernel in another way. Since $K_0(x_0, x_0) = 0$, every function in $\mathcal{H}(K_0)$ vanishes at the point x_0. To see this we use Cauchy–Schwarz. We have $\|\langle f, k_{x_0} \rangle\| \leq \|f\| \|k_{x_0}\| = \|f\| K_0(x_0, x_0)^{1/2} = 0$.

It is possible to show that, in fact, these are precisely the elements of $\mathcal{H}(K_0)$. However, we give an independent proof that will allow us to deduce the matrix positivity condition in Theorem 4.2.

Proof. First we need a Hilbert space result. Let $\mathcal{H}$ be a Hilbert space and let $y \in \mathcal{H}$. Let P denote the projection onto the span of y. The projection P is given by $P(x) = \frac{\langle x, y \rangle}{\|y\|^2} y$. Hence, the projection onto the orthogonal complement of y is given by $P^{\perp}x = (I - P)x = x - \frac{\langle x, y \rangle}{\|y\|^2}$.

Let $\mathcal{H}_0$ be the RKHS space of functions in $\mathcal{H}(K)$ that vanish at x_0 and let P_0 denote the projection onto this subspace. Note that $\mathcal{H}_0 = \{k_{x_0}\}^{\perp}$. Thus, $P_0 = I - P$, where P is the projection onto the span of k_{x_0}. We showed in Theorem 2.5 that the kernel K_0 of the subspace $\mathcal{H}_0$ is given by $\langle P_0(k_y), k_x \rangle$. We now compute

$$\langle P_0(k_y), k_x \rangle = \left\langle k_y - \frac{\langle k_y, k_{x_0} \rangle}{\|k_{x_0}\|^2} k_{x_0}, k_x \right\rangle$$

$$= K(x, y) - \frac{K(x, x_0) K(x_0, y)}{K(x_0, x_0)} = K_0(x, y).$$

This proves that K_0 is a kernel and since it is the kernel for $\mathcal{H}_0$, that $\mathcal{H}(K_0) = \mathcal{H}_0$. $\qquad\square$

We promised that the above result could be used to deduce Cholesky's theorem. To do this, let $X = \{1, \ldots, n\}$ and recall that a positive semidefinite matrix P defines a kernel K on X by setting $K(i, j) = p_{i,j}$. Let $K_0(i, j)$ be the kernel for the space of functions in $\mathcal{H}(K)$ that vanish at $x_0 = k$ given by the above theorem. By the above theorem, $K_0(i, j) = p_{i,j} - p_{i,1} p_{1,j} / p_{k,k}$. Note that these are exactly the entries of the matrix in Cholesky's theorem.

Since K_0 is a kernel function, $(K_0(i, j))$ is a positive semidefinite matrix and Cholesky's theorem follows.

4.2 Schur products and the Schur decomposition

Definition 4.6. Let $A, B \in M_n$. We define the Schur product of $A = (a_{i,j})$ and $B = (b_{i,j})$ by

$$(a_{i,j}) \circ (b_{i,j}) = (a_{i,j} b_{i,j}).$$

Unlike the usual product of matrices we note that the Schur product is commutative. In addition, $(A + B) \circ C = A \circ C + B \circ C$ and $A \circ (B \circ C) = (A \circ B) \circ C$.

The proof of the Cholesky factorization not only provides an algorithm to factor the matrix P in the form LL^*, but we shall see that this algorithm also allows us to decompose the positive matrix P into a sum of rank one positive matrices. We leave it to the exercises (4.1, 4.2) to show that Q is a rank one positive matrix if and only if $Q = vv^*$ for some vector $v \neq 0$. Note that if $v = (\alpha_1, \ldots, \alpha_n)^t$, then $vv^* = (\alpha_i \overline{\alpha_j})$.

Matrices of the form $P = \sum_{k=1}^m v_k v_k^*$ are positive, since

$$\langle Px, x \rangle = \sum_{k=1}^m \langle vv^* x, x \rangle = \sum_{k=1}^m \langle \langle x, v \rangle v, x \rangle = \sum_{k=1}^m |\langle x, v \rangle|^2.$$

The following result shows the converse.

Theorem 4.7 (Schur decomposition). *Let P be an $n \times n$ matrix. If P is positive semidefinite, then there exist vectors $v_1, \ldots, v_n \in \mathbb{C}^n$ such that $P = \sum_{k=1}^n v_k v_k^*$. Moreover, an $n \times n$ matrix of the form $P = \sum_{k=1}^m w_k w_k^*$ is positive definite, and consequently invertible, if and only if the vectors $\{w_k\}_{k=1}^m$ span $\mathbb{C}^n$.*

Proof. If $A, B \in M_n$, the columns of A are the vectors $C_1, \ldots, C_n$, and the rows of B are the vectors $R_1, \ldots, R_n$, then $AB = C_1 R_1 + \cdots + C_n R_n$.

Now apply the Cholesky factorization to the matrix P and write $P = LL^*$, with L lower-triangular. Let $v_1, \ldots, v_n$ be the columns of L. The rows of L^* are the vectors $v_1^*, \ldots, v_n^*$. Hence, $P = \sum_{k=1}^n v_k v_k^*$.

Note that when $P = \sum_{k=1}^m w_k w_k^*$, then the range of P is contained in the span of the vectors $\{w_k\}$. Thus, if P is positive definite, it is invertible, and so we must have that these vectors span $\mathbb{C}^n$.

Conversely, suppose that these vectors span $\mathbb{C}^n$. Recall that $\langle Ph, h \rangle = \sum_{k=1}^n \langle w_k w_k^* h, h \rangle = \sum_{k=1}^n |\langle w_k, h \rangle|^2$. Thus, we see that $\langle Ph, h \rangle = 0$ if and only if $h \perp w_k$ for all k, which is if and only if $h = 0$, since these vectors span $\mathbb{C}^n$. Hence, P is positive definite and, consequently, invertible. $\qquad\square$

Theorem 4.8 (Schur product). *Let P and Q be $n \times n$ matrices. If P and Q are positive semidefinite (respectively, definite), then the Schur product $P \circ Q$ is positive semidefinite (respectively, definite).*

Proof. First we consider the case where P and Q are both rank one matrices. Then by Exercise 4.2 there are vectors $v, w \in \mathbb{C}^n$ such that $P = vv^*$ and $Q = ww^*$. If $v = (v_1, \ldots, v_n)^t$ and $w = (w_1, \ldots, w_n)^t$, then (i, j)-entry of $P \circ Q$ is $(v_i \overline{v_j})(w_i \overline{w_j}) = v_i w_i \overline{v_j w_j}$. Hence $P \circ Q = zz^* \geq 0$ where $z_i = v_i w_i$. In general, we can write $P = \sum_{i=1}^n v_i v_i^*$ and $Q = \sum_{i=1}^n w_i w_i^*$. Using the fact that a sum of rank one positive semidefinite matrices is positive semidefinite and the distributive property of the Schur product, we get

$$P \circ Q = \sum_{i,j=1}^n (v_i v_i^*) \circ (w_j w_j^*) \geq 0,$$

which is a sum of n^2 rank one positive semidefinite matrices.

Now we consider the positive definite case. By Theorem 4.7 the sets of vectors $\{v_i\}$ and $\{w_j\}$ both span $\mathbb{C}^n$. Given a vector v with entries $\alpha_1, \ldots, \alpha_n$ and w with entries $\beta_1, \ldots, \beta_n$, let $v \circ w$ be the vector in $\mathbb{C}^n$ with entries $\alpha_1 \beta_1, \ldots, \alpha_n \beta_n$. Then we have that

$$P \circ Q = \sum_{i,j=1}^n (v_i \circ w_j)(v_i \circ w_j)^*,$$

so that $P \circ Q$ will be positive definite if and only if the n^2 vectors $\{v_i \circ w_j\}_{i,j=1}^n$ span $\mathbb{C}^n$. Finally, note that

$$\langle v \circ w, h \rangle = \langle v, \overline{w} \circ h \rangle.$$

Since the v_is span $\mathbb{C}^n$, we have that $\langle v_i \circ w_j, h \rangle = 0$ for all i, j if and only if $\overline{w_j} \circ h = 0$ for all j. But if this is the case, then letting η be the vector of all ones, we have that

$$0 = \langle \eta, \overline{w_j} \circ h \rangle = \langle \eta \circ w_j, h \rangle = \langle w_j, h \rangle$$

for every j. Since the w_js span $\mathbb{C}^n$, we have that $h = 0$, and so the vectors $v_i \circ w_j$ span $\mathbb{C}^n$. $\qquad\square$

If K_1 and K_2 are kernels on a set X, then we see from Schur's theorem that $K(x, y) = K_1(x, y)K_2(x, y)$ is a kernel on X, since for any finite set of points we have

$$\left(K(x_i, x_j)\right) = \left(K_1(x_i, x_j)K_2(x_i, x_j)\right) = \left(K_1(x_i, x_j)\right) \circ \left(K_2(x_i, x_j)\right) \geq 0.$$

Therefore, Moore's theorem shows that there is an associated RKHS. This RKHS is not as simple to describe as the RKHS associated with the Schur complement. We will examine the RKHS that arises from the product of two kernel functions in detail in the next section.

4.3 Tensor products of Hilbert spaces

We assume that the reader is familiar with the tensor product of vector spaces; for those unfamiliar, see [7]. Suppose that we are given two Hilbert spaces $\mathcal{H}$ and $\mathcal{K}$ with inner products $\langle \cdot, \cdot \rangle_{\mathcal{H}}$ and $\langle \cdot, \cdot \rangle_{\mathcal{K}}$, vectors $h, \hat{h} \in \mathcal{H}$, and $k, \hat{k} \in \mathcal{K}$. Then it is known that setting

$$\left\langle h \otimes k, \hat{h} \otimes \hat{k} \right\rangle = \left\langle h, \hat{h} \right\rangle_{\mathcal{H}} \left\langle k, \hat{k} \right\rangle_{\mathcal{K}},$$

and extending to sums, defines an inner product on the tensor product $\mathcal{H} \otimes \mathcal{K}$, which may then be completed to define a new Hilbert space, generally, still denoted $\mathcal{H} \otimes \mathcal{K}$.

We shall occasionally need to use tensor products in this book, so we gather a few facts here. One reason for covering this material here is that, while it follows from general facts about tensor products that the above definition extends to define a sesquilinear form on the tensor product, it is harder to see that this form satisfies $\langle u, u \rangle \geq 0$ and that $\langle u, u \rangle = 0$ implies $u = 0$.

However, Schur products give us an easy way to see these important facts, so we present proofs here.

Suppose that we are given $h_1, \ldots, h_n \in \mathcal{H}$ and $k_1, \ldots, k_n \in \mathcal{K}$ and $u = \sum_{j=1}^{n} h_j \otimes k_j$. Our first goal is to show that $\langle u, u \rangle \geq 0$. Note that $P = \left(\langle h_j, h_i \rangle_{\mathcal{H}}\right)$ and $Q = \left(\langle k_j, k_i \rangle_{\mathcal{K}}\right)$ are both positive semidefinite matrices. In fact, they are the transposes of the Grammian matrices 2.23. To see this directly, note that if v is the vector in $\mathbb{C}^n$ with entries $\alpha_1, \ldots \alpha_n$, then $\langle Pv, v \rangle = \|\alpha_1 h_1 + \cdots \alpha_n h_n\|^2 \geq 0$, with a similar argument for Q.

Let $\eta \in \mathbb{C}^n$ be the vector of all ones, then

$$\langle u, u \rangle = \sum_{i,j=1}^{n} \langle h_j, h_i \rangle_{\mathcal{H}} \langle k_j, k_i \rangle_{\mathcal{K}} = \langle (P \circ Q)\eta, \eta \rangle_{\mathbb{C}^n} \geq 0,$$

since $P \circ Q$ is positive semidefinite by the Schur product theorem.

Finally, we must show that if $u \neq 0$ then $\langle u, u \rangle > 0$. To do this we use the fact (left as an exercise) that for any tensor product of vector spaces, any element in the tensor product can be expressed as $u = \sum_{i=1}^{n} h_i \otimes k_i$ with $\{h_i\}_{i=1}^{n}$ and $\{k_i\}_{i=1}^{n}$ both linearly independent sets. In this case, using the formula expressing $\langle Pv, v \rangle$ as the norm of a vector, we see that $\langle Pv, v \rangle \neq 0$ when $v \neq 0$. Thus, P and similarly Q are both positive definite. Hence, $P \circ Q$ is positive definite by the Schur product theorem, and so $\langle u, u \rangle = \langle (P \circ Q)\eta, \eta \rangle \neq 0$.

Thus, we have shown that the sesquilinear form defined on the tensor product is really an inner product.

Another fact that we will sometimes need about tensor products concerns orthonormal bases.

Proposition 4.9. *Let $\mathcal{H} \otimes \mathcal{K}$ be the tensor product of two Hilbert spaces. If $\{e_s\}_{s \in S}$ and $\{f_t\}_{t \in T}$ are orthonormal bases for $\mathcal{H}$ and $\mathcal{K}$, respectively, then $\{e_s \otimes f_t\}_{(s,t) \in S \times T}$ is an orthonormal basis for $\mathcal{H} \otimes \mathcal{K}$.*

Proof. It is easily checked that these vectors form an orthonormal set. To prove that they are an orthonormal basis, it will suffice to show that their linear span is dense in $\mathcal{H} \otimes \mathcal{K}$.

To prove this it will be enough to show that every vector of the form $h \otimes k$ is in their closed linear span. Since $\{e_s\}_{s \in S}$ is an orthonormal basis for $\mathcal{H}$, we can choose a sequence of vectors $\{h_n\}$ in their span with $\|h - h_n\| \to 0$, and similarly, we may choose a sequence of vectors $\{k_n\}$ in the span of $\{f_t\}_{t \in T}$ such that $\|k - k_n\| \to 0$. Then $h_n \otimes k_n$ is in the linear span of $\{e_s \otimes f_t\}$ and

$$\|h \otimes k - h_n \otimes k_n\| = \|(h - h_n) \otimes k + h_n \otimes (k - k_n)\| \leq$$
$$\|h - h_n\|\|k\| + \|h_n\|\|k - k_n\| \to 0.$$

Thus, $h \otimes k$ is in the closed linear span of $\{e_s \otimes f_t\}$ and the proof is complete. $\qquad\square$

4.4 Kernels arising from polynomials and power series

The Schur product theorem gives us an important tool for generating many important kernel functions. Below we look at several such families. Although all of these could be merged into the most general example, we start with the simplest examples first.

4.4.1 Polynomial kernels

Let $p(x) = \sum_{n=0}^{N} a_n x^n$ be a polynomial with $a_n \geq 0$ for all n. We claim that the function

$$K_p : \mathbb{C} \times \mathbb{C} \to \mathbb{C}$$

defined by $K_p(z, w) = p(z\overline{w}) = \sum_{n=0}^{N} a_n(z\overline{w})^n$ is a kernel function.

To see this, choose points $z_1, \ldots, z_m \in \mathbb{C}$ and let Q be the $m \times m$ rank one positive matrix $Q = (z_i \overline{z_j})$. Then the $m \times m$ matrix $(K_p(z_i, z_j))$ satisfies

$$(K_p(z_i, z_j)) = a_0 J_m + a_1 Q + a_2 Q \circ Q + \cdots + a_N Q^{\circ N},$$

where J_m is the matrix of all ones and the last term denotes the Schur product of Q with itself N times. Since $a_j \geq 0$, $J_m \geq 0$ and $Q \geq 0$, by the Schur product theorem each term $a_j Q^{\circ j}$ is positive semidefinite and, hence, the sum is positive semidefinite.

Thus, K_p is a kernel function on $\mathbb{C}$ and by Moore's theorem there is an RKHS $\mathcal{H}(K_p)$ of functions on $\mathbb{C}$ such that, for each $w \in \mathbb{C}$, the kernel function is the polynomial of degree at most N given by $k_w(z) = p(z\overline{w})$.

If we set $f_n(z) = \sqrt{a_n} z^n$, then $K(z, w) = \sum_{n=0}^{N} f_n(z) \overline{f_n(w)}$. By Exercise 3.7 and Exercise 2.9, the functions $f_0, \ldots, f_N$ form a Parseval frame for the space. Hence, $\mathcal{H}(K_p) = \text{span}\{f_0, \ldots, f_N\}$ and we see that $\mathcal{H}(K_p)$ is a vector space of polynomials of degree at most N. Moreover, if we consider only those n for which $a_n \neq 0$, then the corresponding set of f_ns is linearly independent, and hence is an orthonormal basis for $\mathcal{H}(K_p)$, by Exercise 2.9. That is, $\mathcal{H}(K_p) = \text{span}\{f_j : a_j \neq 0\}$ and $z^k \notin \mathcal{H}(K_p)$ whenever $a_k = 0$.

4.4.2 Power series on disks

There are two important generalizations of polynomial kernels. Let $p(x) = \sum_{n=0}^{+\infty} a_n x^n$ be a power series with $a_n \geq 0$. Suppose that the radius of convergence of this power series is $r^2 > 0$. We show that such a series can be used to define an RKHS of analytic functions on $\mathbb{D}_r$ the open disk of radius r. In the next section we examine how this same construction yields spaces in several complex variables.

In parallel with the previous example, we can use p to define a kernel function on $K_p : \mathbb{D}_r \times \mathbb{D}_r \to \mathbb{C}$ by setting

$$K_p(z, w) = \sum_{n=0}^{\infty} a_n(z\overline{w})^n.$$

Since $z, w \in \mathbb{D}_r$ implies that $|z\overline{w}| < r^2$, this power series will converge.

If $z_1, \ldots, z_m \in \mathbb{D}_r$, then as before the $m \times m$ matrix $Q = (z_i \overline{z_j})$ is positive semidefinite and the matrix $(K_p(z_i, z_j))$ can be seen to be the entrywise limit of the matrices $P_N = \sum_{n=0}^{N} a_n Q^{\circ n}$, where $Q^{\circ 0}$ denotes the all 1 matrix J_m. Since each of the matrices P_N is positive semidefinite, the limit matrix $(K_p(z_i, z_j))$ will also be positive semidefinite. Thus, K_p defines an RKHS of functions on $\mathbb{D}_r$.

If we set $f_n(z) = \sqrt{a_n} z^n$, then on $\mathbb{D}_r \times \mathbb{D}_r$ we have that $K_p(z, w) = \sum_{n=0}^{\infty} f_n(z) \overline{f_n(w)}$ where the sum is the pointwise limit of the partial sums. Thus, arguing as above, the functions $f_n(z)$ for which $a_n \neq 0$ will be a Parseval frame for the RKHS $\mathcal{H}(K_p)$ and when $a_k = 0$, then $z^k \notin \mathcal{H}(K_p)$.

We would like to show that these functions also form an orthonormal basis for the space, but since we are now dealing with an infinite dimensional space we can no longer appeal to Exercise 2.9 and we need a subtler notion of linear independence.

Definition 4.10. Let $\{f_s : s \in S\}$ be a set of vectors in the Hilbert space $\mathcal{H}$. We say that these vectors are $\boxed{\ell^2\text{-INDEPENDENT}}$, provided that whenever $\alpha_s \in \mathbb{C}$ satisfy $\sum_{s \in S} |\alpha_s|^2 < \infty$ and the sum $\sum_{s \in S} \alpha_s f_s$ converges in norm to 0, then $\alpha_s = 0$ for every $s \in S$.

Proposition 4.11. *Let* $\{f_s : s \in S\}$ *be a Parseval frame for the Hilbert space* $\mathcal{H}$. *Then* $\{f_s : s \in S\}$ *is* ℓ^2-*independent if and only if* $\{f_s : s \in S\}$ *is an orthonormal basis for* $\mathcal{H}$.

Proof. Clearly, an orthonormal basis is ℓ^2-independent.

Conversely, since the set is a Parseval frame, the map $V : \mathcal{H} \to \ell^2(S)$ given by $V(h) = (\langle h, f_s \rangle)$ is an isometry. We've seen that the adjoint of this map, $V^* : \ell^2(S) \to \mathcal{H}$ has the property that $V^*(e_s) = f_s$ where $\{e_s : s \in S\}$ is the canonical orthonormal basis for $\ell^2(S)$, i.e. e_s is 1 in the s-th entry and 0 in all other entries. Thus, the statement that $\{f_s : s \in S\}$ is ℓ^2-independent is equivalent to the statement that V^* is one-to-one.

Since $\mathcal{N}(V^*) = \mathcal{R}(V)^{\perp}$ and the range of V is closed. We have that V^* is one-to-one if and only if V is an onto isometry, i.e. a unitary, in which case V^* is also a unitary. But the image of an orthonormal basis under a unitary transformation is also orthonormal and so the result follows. $\square$

We summarize:

Theorem 4.12. *Let* $p(x) = \sum_{n=0}^{+\infty} a_n x^n$, $a_n \geq 0$ *be a power series with radius of convergence* $r^2 > 0$ *and set* $K_p(z, w) = \sum_{n=0}^{+\infty} a_n (z \overline{w})^n$. *Then* K_p *is a kernel for an RKHS of analytic functions on* $\mathbb{D}_r$ *and the set of functions* $f_n(z) = \sqrt{a_n} z^n$ *for which* $a_n \neq 0$ *is an orthonormal basis for* $\mathcal{H}(K_p)$.

Proof. All that remains to show is that this set is an orthonormal basis for $\mathcal{H}(K_p)$. By Papadakis's theorem, we know that it is a Parseval frame for $\mathcal{H}(K_p)$, so by the above it will be enough to show that it is ℓ^2-independent.

Let $S = \{n : a_n \neq 0\}$. If $\sum_{n \in S} \alpha_n f_n$ converges to an element of $\mathcal{H}(K_p)$ then that element is an analytic function on $\mathbb{D}^r$. The power series defines the 0 function if and only if each of its coefficients is 0. Hence, if the sum converges to 0 then $\alpha_n \sqrt{a_n} = 0$ for every $n \in S$. But this implies that $\alpha_n = 0$ for every $n \in S$, and so the set is orthonormal. $\qquad\qquad\qquad\qquad\qquad\qquad\qquad\square$

Some important applications of this theorem include the case that $a_n = 1$ for all n, in which case $p(x) = \frac{1}{1-x}$ has radius of convergence 1, so that $K_p(z, w) = \frac{1}{1-z\overline{w}}$ is the $\boxed{\text{SZEGÖ KERNEL}}$ and the RKHS that we obtain is the $\boxed{\text{HARDY SPACE ON THE DISK } H^2(\mathbb{D})}$, which was one of our earliest examples. When $a_n = n + 1$, $p(x) = \frac{1}{(1-x)^2}$, so that $K_p(z, w) = \frac{1}{(1-z\overline{w})^2}$ is the $\boxed{\text{BERGMAN KERNEL}}$ and we recover the *Bergman space* $B^2(\mathbb{D})$.

Another important example is when $a_n = \frac{1}{n!}$, so that $p(x) = e^x$ has infinite radius of convergence and

$$K_p(z, w) = e^{z\overline{w}}$$

is defined on the whole complex plane. In this case we obtain a space of functions that are analytic on the whole complex plane, i.e. $\boxed{\text{ENTIRE FUNCTIONS}}$. This kernel is called the $\boxed{\text{SEGAL–BARGMANN KERNEL}}$ and the RKHS is known as the $\boxed{\text{SEGAL–BARGMANN SPACE}}$.

Note that in each of these cases, since we have that the set of functions $f_n(z) = \sqrt{a_n} z^n$, such that $a_n \neq 0$, are an orthonormal basis, we have that $\|z^n\| = \frac{1}{\sqrt{a_n}}$. For this reason, a space that is obtained as $\mathcal{H}(K_p)$, for some power series p, as above, is referred to as a $\boxed{\text{WEIGHTED HARDY SPACE}}$ and the numbers $\sqrt{a_n}$ (or their reciprocals) are referred to as the weights. Some authors require that $a_n \neq 0$ for all n to be a weighted Hardy space, while other authors do not even require the power series to have a nonzero radius of convergence! Weighted Hardy spaces are also referred to as $\boxed{\text{SHIELD'S } H^2_\beta \text{ SPACES}}$, especially in the case of 0 radius of convergence.

Now we turn our attention to some examples of RKHSs whose domains are higher dimensional sets, such as a ball in $\mathbb{C}^M$.

4.4.3 Power series on balls

The simplest such example is when we start with a Hilbert space $\mathcal{L}$, and look at the set of bounded, linear functionals on $\mathcal{L}$. In Proposition 2.24, we saw that this was an RKHS with kernel $K(x, y) = \langle x, y \rangle$. Our results on Schur products show us that for each positive integer N the function $\langle x, y \rangle^N$ also defines a kernel on $\mathcal{L}$ and hence an RKHS of functions on $\mathcal{L}$. In this section, we examine this space and then use it as a building block for determining various spaces of power series on balls in $\mathcal{L}$.

Proposition 4.13. *Let $\mathcal{L}$ be a Hilbert space, let N be a positive integer and define $K_N : \mathcal{L} \times \mathcal{L} \to \mathbb{C}$ by $K_N(x, y) = \langle x, y \rangle^N$. Then $\mathcal{H}(K_N)$ contains all functions $g : \mathcal{L} \to \mathbb{C}$ that can be written as a linear combination of products of N bounded linear functionals.*

Proof. Since $\mathcal{H}(K_N)$ is a vector space, it will be enough to prove that if $w_1, \ldots, w_N \in \mathcal{L}$ and

$$g(x) = \langle x, w_1 \rangle \cdots \langle x, w_N \rangle,$$

then $g \in \mathcal{H}(K_N)$.

By the above calculation, we know that $f_i(x) = \langle x, w_i \rangle \in \mathcal{H}(K_1)$. Hence, by Theorem 3.11 there exist constants C_i so that $0 \le f_i(x)\overline{f_i(y)} \le C_i^2 K(x, y)$ in the ordering on kernel functions. Now note (Exercise 4.5) that if $0 \le P_i \le Q_i$ are positive semidefinite matrices, then $0 \le P_1 \circ P_2 \le Q_1 \circ Q_2$.

Thus, we have that

$$0 \le g(x)\overline{g(y)} = \prod_{i=1}^{N} f_i(x)\overline{f_i(y)} \le C^2 K_1(x, y)^N,$$

where $C = C_1 \cdots C_N$. Applying Theorem 3.11 again implies that $g \in \mathcal{H}(K_N)$. $\square$

This result seems a bit abstract, but it becomes much more concrete when we consider the case of $\mathcal{L} = \mathbb{C}^M$ with the usual inner product and denote vectors in this space by $z = (z_1, \ldots, z_M)$ and $w = (\overline{a_1}, \ldots, \overline{a_M})$. Then a linear functional

$$f_w(v) = a_1 z_1 + \cdots a_M z_M$$

is just a homogeneous first order polynomial in the coordinate variables $z_1, \ldots, z_M$. Note that the kernel function for the point w is the function

$$K_N(z, w) = (a_1 z_1 + \cdots + a_M z_M)^N.$$

A product of N such linear functionals is just a homogeneous polynomial of degree N. In particular, if we let $w = e_i$, $1 \le i \le M$ denote the canonical orthonormal basis for $\mathbb{C}^M$, then $f_{e_i}(z) = z_i$. Thus, by taking products of N such linear functions we can get any monomial of the form

$$z_1^{j_1} \cdots z_M^{j_M} \text{ with } j_1 + \cdots j_M = N.$$

In this case, the number N is called the $\boxed{\text{TOTAL DEGREE}}$ of the monomial.

Thus, the above result says that for $\mathcal{L} = \mathbb{C}^M$ and $K_N(x, y) = \langle x, y \rangle^N$, then $\mathcal{H}(K_N)$ contains the vector space of homogeneous polynomials of total degree N.

In the case that $\mathcal{L}$ is finite dimensional, one can show that $\mathcal{H}(K_N)$ is exactly the set of homogeneous polynomials (Exercise 4.10). However, when $\mathcal{L}$ is infinite dimensional, then $\mathcal{H}(K_N)$ is larger.

Computing the norm of a homogeneous polynomial in this space can be done by appealing to Theorem 3.11, but can be a tricky business. In Chapter 7 we will see another way to compute these norms and give a more effective way of characterizing the functions that are in these spaces.

Theorem 4.14. *Let $p(x) = \sum_{n=0}^{\infty} a_n x^n$ be a power series with $a_n \geq 0$ for all n and radius of convergence $r^2 > 0$, let $\mathcal{L}$ be a Hilbert space and let $\mathbb{B}_r$ denote the ball of radius r in $\mathcal{L}$. If we define $K_p : \mathbb{B}_r \times \mathbb{B}_r \to \mathbb{C}$ by $K_p(x, y) = p(\langle x, y \rangle)$, then K_p is a kernel on $\mathbb{B}_r$. For each n such that $a_n > 0$, $\mathcal{H}(K_p)$ contains every function on $\mathbb{B}_r$ that can be written as a linear combination of products of n bounded linear functionals.*

Proof. Let $p_N(x) = \sum_{n=0}^{N} a_n x^n$, then for any pair of vectors $x, y \in \mathbb{B}_r$, $\lim_{N \to \infty} p_N(\langle x, y \rangle) = p(\langle x, y \rangle)$. Hence, for any set of vectors $x_1, \dots, x_n$ in $\mathbb{B}_r$ the $n \times n$ matrix $\big(K_p(x_i, x_j) \big)$ will be the entrywise limit of the matrices $\big(K_{p_N}(x_i, x_j) \big)$. Thus, to prove that K_p is a kernel function, it will be enough to prove that each K_{p_N} is a kernel function.

Let $K_J(x, y) = \langle x, y \rangle^J$, then we know that K_J is a kernel function and since $a_J \geq 0$, we have that $a_J K_J$ is a kernel function. Since sums of kernel functions are kernel functions, we have that K_{p_N} is a kernel function.

Also, when $a_J > 0$ then as sets $\mathcal{H}(K_J) = \mathcal{H}(a_J K_J)$ and since $a_J K_J \leq K_p$ in the ordering on kernels, we have that $\mathcal{H}(K_J) \subseteq \mathcal{H}(K_p)$. Thus, by the previous result when $a_J > 0$, $\mathcal{H}(K_p)$ contains every function on $\mathcal{L}$ that can be written as a product of J bounded linear functionals and the result follows. $\square$

Thus, in the case that $\mathcal{L} = \mathbb{C}^M$ we see that for each N such that $a_N > 0$, then $\mathcal{H}(K_p)$ contains every homogeneous polynomial of total degree N.

4.4.4 The Drury–Arveson space

If we consider the case where $a_n = 1$ for all n, then $p(x) = \sum_{n=0}^{\infty} x^n = \frac{1}{1-x}$ has radius of convergence $r = 1$.

For a general Hilbert space $\mathcal{L}$, we obtain a kernel function on the unit ball of $\mathcal{L}$ defined by

$$K_p(v, w) = \sum_{n=0}^{\infty} \langle v, w \rangle^n = \frac{1}{1 - \langle v, w \rangle}.$$

This kernel is called the *Drury–Arveson kernel* and the corresponding RKHS on the unit ball is called the *Drury–Arveson space*.

In particular, when $\mathcal{L} = \mathbb{C}$, the Drury–Arveson kernel is the Szegö kernel and the Drury–Arveson space is the Hardy space. Thus, the Drury–Arveson space is a natural generalization of the Hardy space to an RKHS on the unit ball of higher dimensional Hilbert spaces.

We now wish to say a little about the functions that appear in this space. In Chapter 7 we will find another characterization of this space that gives a complete description of the functions in this space and a better method to compute their norm.

Proposition 4.15. *Let $\mathcal{L}$ be a Hilbert space and let $\mathbb{B}_1 \subset \mathcal{L}$ denote its unit ball. Then the Drury–Arveson space on $\mathbb{B}_1$ contains every function that can be expressed as a sum of products of finitely many bounded linear functionals. In particular, for $\mathcal{L} = \mathbb{C}^M$ the Drury–Arveson space contains every polynomial in the coordinate functions.*

Proof. Note that $a_n > 0$ for all n and apply Theorem 4.14. $\qquad\square$

The Drury–Arveson space has been the subject of considerable study in multivariable complex analysis and operator theory for the last thirty years. We will see later that it also plays a role in the chapter on statistics.

From this viewpoint, the Drury–Arveson space is the "natural" multivariable generalization of the Hardy space. However, earlier researchers considered the space $H^2(\mathbb{D}^N)$ to be the "natural" generalization of $H^2(\mathbb{D})$ and, consequently, it is called *the multivariable Hardy space*.

4.4.5 The Segal–Bargmann space

For another example of the above construction we return to the exponential function. If we set $a_n = \frac{1}{n!}$ so that $p(x) = e^x$, which has an infinite radius of convergence, then we saw that

$$K_p(z, w) = e^{z\overline{w}}$$

defines a kernel on the complex plane which we called the *Segal–Bargmann kernel*.

Thus, for any Hilbert space $\mathcal{L}$, setting

$$K_p(v, w) = \sum_{n=0}^{\infty} \frac{1}{n!} \langle v, w \rangle^n = e^{\langle v, w \rangle}$$

defines an RKHS of functions on $X = \mathcal{L}$. This kernel is called the
SEGAL–BARGMANN KERNEL ON $\mathcal{L}$ and the resulting RKHS is called the
SEGAL–BARGMANN SPACE ON $\mathcal{L}$. As with the Drury–Arveson space this
space will contain any function that can be written as a sum of products of
bounded linear functionals.

If we replace $\mathbb{C}$ by $\mathbb{C}^M$ and denote points by $z = (z_1, \ldots, z_M)$ and $w = (w_1, \ldots, w_M)$, then we obtain the *Segal–Bargmann kernel on $\mathbb{C}^M$* which is
given by the formula

$$K_p(z, w) = e^{\langle z, w \rangle} = e^{z_1 \overline{w_1}} \cdots e^{z_N \overline{w_M}},$$

which is just the product of the 1-dimensional Segal–Bargmann kernels.

4.4.6 Other power series

More generally, one can consider kernels that arise as the composition of a
power series with a kernel function. The following is an example of one such
result. We leave its proof as an exercise.

Theorem 4.16. *Let $p(x) = \sum_{n=0}^{\infty} a_n x^n$, $a_n \geq 0$ be a power series with an
infinite radius of convergence, let X be a set and let $K : X \times X \to \mathbb{C}$ be a
kernel function. Then the function $p \circ K : X \times X \to \mathbb{C}$ defined by $p \circ K(x, y) = p(K(x, y))$ is a kernel function. For each n such that $a_n > 0$, $\mathcal{H}(p \circ K)$
contains every function that can be written as a linear combination of products
of n functions in $\mathcal{H}(K)$.*

Thus, in particular if $K : X \times X \to \mathbb{C}$ is a kernel, then e^K is a kernel
and $\mathcal{H}(e^K)$ contains every function that is in the algebra of functions on X
generated by $\mathcal{H}(K)$.

4.5 Exercises

Exercise 4.1. Prove that an $n \times n$ matrix is rank one if and only if it is of the
form $(\alpha_i \overline{\beta_j})$ for some choice of scalars.

Exercise 4.2. Prove that an $n \times n$ rank one matrix is positive if and only if it
is of the form $(\alpha_i \overline{\alpha_j})$ for some choice of scalars.

Exercise 4.3. Prove that every positive $n \times n$ rank k matrix can be written as a sum of k positive rank one matrices.

Exercise 4.4. Prove that the Schur product is distributive over sums, i.e. $(A + B) \circ (C + D) = A \circ C + B \circ C + A \circ D + B \circ D$.

Exercise 4.5. Prove that if $0 \leq P_i \leq Q_i$ for $i = 1, 2$, then $0 \leq P_1 \circ P_2 \leq Q_1 \circ Q_2$.

Exercise 4.6. Let $K_i : X \times X \to \mathbb{C}, i = 1, 2$ be kernel functions and let $K(x, y) = K_1(x, y)K_2(x, y)$. Prove that if $f_i \in \mathcal{H}(K_i), i = 1, 2$, then $F(x) = f_1(x)f_2(x) \in \mathcal{H}(K)$.

Exercise 4.7. Let $\mathcal{H}$ and $\mathcal{K}$ be vector spaces and let $u \in \mathcal{H} \otimes \mathcal{K}$. The *rank of* u, denoted rank(u) is defined to be the smallest number n of vectors needed to write $u = \sum_{i=1}^{n} h_i \otimes k_i$. Note that this minimum always exists. Prove that when we write $u = \sum_{i=1}^{n} h_i \otimes k_i$ with $n = rank(u)$, then $\{h_i\}_{i=1}^{n}$ and $\{k_i\}_{i=1}^{n}$ are both linearly independent sets of vectors.

Exercise 4.8. Let $\mathcal{L}$ be a Hilbert space, let $w \in \mathcal{L}$, let $K(x, y) = \langle x, y \rangle$ and let $f(x) = \langle x, w \rangle$. Prove that

$$f(x)\overline{f(y)} \leq \|w\|^2 K(x, y)$$

in the ordering on kernel functions and that $\|w\|^2$ is the least such constant.

Exercise 4.9. Show that the power series $\sum_{n=1}^{\infty} \frac{x^n}{n}$ has radius of convergence 1 and converges to $-\ln(1 - x)$. Conclude that the function

$$K(z, w) - 1 - \ln(1 - z\overline{w}) = 1 + \ln(\frac{1}{1 - z\overline{w}})$$

defines a kernel function on $\mathbb{D}$. This kernel is called the $\boxed{\text{DIRICHLET KERNEL}}$ and the resulting RKHS is called the $\boxed{\text{DIRICHLET SPACE}}$ on the disk. Prove that $f(z) = \sum_{n=0}^{\infty} b_n z^n$ is in the Dirichlet space if and only if $\sum_{n=1}^{\infty} n|b_n|^2 < +\infty$. Prove that if f is in the Dirichlet space if and only if f' is in the Hardy space.

Exercise 4.10. Let $K_N : \mathbb{C}^M \times \mathbb{C}^M \to \mathbb{C}$ be defined by $K_N(z, w) = \langle z, w \rangle^N$ with $z = (z_1, \ldots, z_M)$. Prove that $\mathcal{H}(K_N)$ is the set of homogeneous polynomials of total degree N. [Hint: use the fact that finite dimensional normed spaces are automatically complete.]

Exercise 4.11. Let $K : \mathbb{C}^M \times \mathbb{C}^M \to \mathbb{C}$ be defined by

$$K(z, w) = \sum_{n=0}^{N} \langle v, w \rangle^n .$$

Prove that K is a kernel and that if we write $v = (z_1, \ldots, z_M)$ then $\mathcal{H}(K)$ is the set of all polynomials in the variables $z_1, \ldots, z_M$ of total degree at most N. (The $\boxed{\text{TOTAL DEGREE}}$ of a monomial $z_1^{j_1} \cdots z_M^{j_M}$ is $j_1 + \cdots + j_M$ and the total degree of a polynomial is the maximum of the total degrees of monomials with non-zero coefficients.)

Exercise 4.12. Prove Theorem 4.16.

Exercise 4.13. The purpose of this exercise is to show that the hypothesis of ℓ^2-independence is really needed for a Parseval frame to be an orthonormal basis. To this end let $\mathcal{H} = \ell^2(\mathbb{N})$ as vector spaces, but we will endow $\mathcal{H}$ with a new inner product. Set $v = (1, 1/2, 1/3, \ldots) \in \ell^2(\mathbb{N})$, let $\langle, \rangle$ denote the usual inner product on $\ell^2(\mathbb{N})$ and define:

$$\langle x, y \rangle_1 = \langle x, y \rangle + \langle x, v \rangle \langle v, y \rangle.$$

1. Prove that $\langle, \rangle_1$ is an inner product on $\mathcal{H}$ and that with respect to this inner product, $\|x\|_1^2 = \|x\|^2 + |\langle x, v \rangle|^2$.
2. Prove that $\mathcal{H}$ is complete in this inner product and hence a Hilbert space.
3. Prove that $\{v, e_1, e_2, \ldots\}$ is a Parseval frame for $\mathcal{H}$.
4. Show that $\{v, e_1, \ldots\}$ is not an orthonormal set in $\mathcal{H}$.
5. Show that $\{v, e_1, \ldots\}$ is a linearly independent set.

5

Operations on kernels

In this chapter we consider how various algebraic operations on kernels affect the corresponding Hilbert spaces. The idea of examining and exploiting such relations, along with many of the results of this chapter, can be traced back to the seminal work of Aronszajn [2].

5.1 Complexification

Let $\mathcal{H}$ be an RKHS of real-valued functions on the set X with reproducing kernel K. Let $\mathcal{W} = \{f_1 + if_2 : f_1, f_2 \in \mathcal{H}\}$, which is a vector space of complex-valued functions on X. If we set

$$\langle f_1 + if_2, g_1 + ig_2 \rangle_{\mathcal{W}} = \langle f_1, g_1 \rangle_{\mathcal{H}} + i\langle f_2, g_1 \rangle_{\mathcal{H}} - i\langle f_1, g_2 \rangle_{\mathcal{H}} + \langle f_2, g_2 \rangle_{\mathcal{H}},$$

then it is easily checked that this defines an inner product on $\mathcal{W}$, with corresponding norm,

$$\|f_1 + if_2\|_{\mathcal{W}}^2 = \|f_1\|_{\mathcal{H}}^2 + \|f_2\|_{\mathcal{H}}^2.$$

Hence, $\mathcal{W}$ is a Hilbert space and since

$$f_1(y) + if_2(y) = \langle f_1 + if_2, k_y \rangle_{\mathcal{W}},$$

we have that $\mathcal{W}$ equipped with this inner product is an RKHS of complex-valued functions on X with the same reproducing kernel, $K(x, y)$.

We call $\mathcal{W}$ the $\boxed{\text{COMPLEXIFICATION OF } \mathcal{H}}$. Since every real-valued RKHS can be complexified in a way that still preserves the reproducing kernel, we shall from this point on primarily consider the case of complex-valued reproducing kernel Hilbert spaces.

5.2 Differences and sums

The first result characterizes inclusions of spaces. This then leads to a characterization of when the difference of two reproducing kernels is again a kernel. Various statements of these results can all be found in Aronszajn's fundamental paper on kernels.

Theorem 5.1 (Aronszajn's inclusion theorem). *Let X be a set and let $K_i :$ $X \times X \to \mathbb{C}, i = 1, 2$ be kernels on X. Then $\mathcal{H}(K_1) \subseteq \mathcal{H}(K_2)$ if and only if there exists a constant $c > 0$ such that $K_1 \le c^2 K_2$. Moreover, $\|f\|_2 \le c \|f\|_1$ for all $f \in \mathcal{H}(K_1)$.*

Proof. First, assume that such a constant $c > 0$ exists. If $f \in \mathcal{H}(K_1)$ with $\|f\|_1 = 1$, then $f(x)\overline{f(y)} \le K_1(x, y) \le c^2 K_2(x, y)$, which implies that $f \in \mathcal{H}(K_2)$ and $\|f\|_2 \le c$. Hence, $\mathcal{H}(K_1) \subseteq \mathcal{H}(K_2)$ and $\|f\|_2 \le c \|f\|_1$.

We now prove the converse. Assume that $\mathcal{H}(K_1) \subseteq \mathcal{H}(K_2)$ and let $T :$ $\mathcal{H}(K_1) \to \mathcal{H}(K_2)$ be the inclusion map, $T(f) = f$. We claim that T is bounded. Let $\{f_n\}$ be a sequence in $\mathcal{H}(K_1)$ that converges to $f \in \mathcal{H}(K_1)$. Now suppose that $g \in \mathcal{H}(K_2)$ with $\|T(f_n) - g\|_2 \to 0$. We have that $f(x) = \lim_n f_n(x) = \lim_n T(f_n)(x) = g(x)$. Hence, $g = T(f)$ and, by the closed graph theorem, T is bounded. We have $\|f\|_2 = \|T(f)\| \le \|T\| \|f\|_1$ for all $f \in \mathcal{H}(K_1)$.

We will show that the constant c can be chosen as the norm of T, i.e. that $K_1 \le c^2 K_2$ for $c = \|T\|$. Let $\{x_1, \dots, x_n\} \subseteq X$, and let $\alpha_1, \dots, \alpha_n \in \mathbb{C}$. We set $k_y^1(x) = K_1(x, y), k_y^2(x) = K_2(x, y)$ and compute:

$$
\begin{aligned}
\sum_{i,j=1}^{n} \overline{\alpha_i} \alpha_j K_1(x_i, x_j) &= \sum_{i,j=1}^{n} \overline{\alpha_i} \alpha_j k_{x_j}^1(x_i) \\
&= \sum_{i,j=1}^{n} \overline{\alpha_i} \alpha_j \left\langle k_{x_j}^1, k_{x_i}^2 \right\rangle_2 \\
&= \left\langle \sum_{j=1}^{n} \alpha_j k_{x_j}^1, \sum_{i=1}^{n} \alpha_i k_{x_i}^2 \right\rangle_2 \\
&\le \left\| \sum_{j=1}^{n} \alpha_j k_{x_j}^1 \right\|_2 \left\| \sum_{i=1}^{n} \alpha_i k_{x_i}^2 \right\|_2 \\
&\le c \left\| \sum_{j=1}^{n} \alpha_j k_{x_j}^1 \right\|_1 \left\| \sum_{i=1}^{n} \alpha_i k_{x_i}^2 \right\|_2 .
\end{aligned}
$$

Let $B = \sum_{i,j=1}^{n} \overline{\alpha_i} \alpha_j K_1(x_i, x_j) = \left\| \sum_{j=1}^{n} \alpha_j k_{x_j}^1 \right\|_1^2$. If we square both sides of the above inequality we get $B^2 \le c^2 B \left(\sum_{i,j=1}^{n} \overline{\alpha_i} \alpha_j K_2(x_i, x_j) \right)$. Cancelling the factor of B gives $K_1 \le c^2 K_2$. $\qquad\square$

Definition 5.2. Given two Hilbert spaces, $\mathcal{H}_i$ with norms, $\|\cdot\|_i$, $i = 1, 2$, we say that $\mathcal{H}_1$ is $\boxed{\text{CONTRACTIVELY CONTAINED}}$ in $\mathcal{H}_2$ provided that $\mathcal{H}_1$ is a (not necessarily closed) subspace of $\mathcal{H}_2$ and for every $h \in \mathcal{H}_1$, $\|h\|_2 \le \|h\|_1$.

Corollary 5.3 (Aronszajn's differences of kernels theorem). *Let $\mathcal{H}_i$, $i = 1, 2$ be RKHSs on the set X with reproducing kernels, K_i, $i = 1, 2$, respectively. Then $\mathcal{H}_1$ is contractively contained in $\mathcal{H}_2$ if and only if $K_2 - K_1$ is a kernel function.*

$\boxed{\text{DE BRANGES SPACES}}$ are a family of reproducing kernel Hilbert spaces on the unit disk that are contractively contained in the Hardy space $H^2(\mathbb{D})$. Properties of de Branges spaces played a key role in L. de Branges's solution of the Bieberbach conjecture [6]. These spaces were studied extensively in the book [5].

If A and B are positive matrices then so is $A + B$. Thus, if K_1 and K_2 are kernel functions on a set X, then so is the function $K = K_1 + K_2$. The next result examines the relationship between the three corresponding RKHSs.

Theorem 5.4 (Aronszajn's sums of kernels theorem). *Let $\mathcal{H}_1$ and $\mathcal{H}_2$ be RKHSs on X with reproducing kernels K_1 and K_2. Let $\|\cdot\|_1$ and $\|\cdot\|_2$ denote the corresponding norms. If $K = K_1 + K_2$, then*

$$\mathcal{H}(K) = \{f_1 + f_2 : f_i \in \mathcal{H}_i, i = 1, 2\}.$$

If $\|\cdot\|$ denotes the norm on $\mathcal{H}(K)$, then for $f \in \mathcal{H}(K)$ we have

$$\|f\|^2 = \min\{\|f_1\|_1^2 + \|f_2\|_2^2 : f = f_1 + f_2, f_i \in \mathcal{H}_i, i = 1, 2\}.$$

Proof. Consider the orthogonal direct sum of the two Hilbert spaces, $\mathcal{H}_1 \oplus \mathcal{H}_2 = \{(f_1, f_2) : f_i \in \mathcal{H}_i\}$ with inner product

$$\langle (f_1, f_2), (g_1, g_2) \rangle = \langle f_1, g_1 \rangle_1 + \langle f_2, g_2 \rangle_2,$$

where $\langle \cdot, \cdot \rangle_i$ denotes the inner product in the Hilbert space, $\mathcal{H}_i, i = 1, 2$. Note that with this inner product $\|(f_1, f_2)\|_{\mathcal{H}_1 \oplus \mathcal{H}_2}^2 = \|f_1\|_1^2 + \|f_2\|_2^2$. Since $\mathcal{H}_i, i = 1, 2$ are both subspaces of the vector space of all functions on X, the intersection, $F_0 = \mathcal{H}_1 \cap \mathcal{H}_2$, is a well-defined vector space of functions on X. Let $\mathcal{N} = \{(f, -f) : f \in F_0\} \subseteq \mathcal{H}_1 \oplus \mathcal{H}_2$.

Note that $\mathcal{N}$ is a closed subspace, since if $\|(f_n, -f_n) - (f, g)\|_{\mathcal{H}_1 \oplus \mathcal{H}_2} \to 0$, then $\|f_n - f\|_1 \to 0$ and $\|-f_n - g\|_2 \to 0$, and hence, at each point, $f(x) = -g(x)$.

Therefore, decomposing $\mathcal{H}_1 \oplus \mathcal{H}_2 = \mathcal{N} + \mathcal{N}^\perp$, we see that every pair, $(f_1, f_2) = (f, -f) + (h_1, h_2)$ with $f \in F_0$ and $(h_1, h_2) \perp \mathcal{N}$.

Let $\mathcal{H}$ denote the vector space of functions of the form $\{f_1 + f_2 : f_i \in \mathcal{H}_i, i = 1, 2\}$ and define $\Gamma : \mathcal{H}_1 \oplus \mathcal{H}_2 \to \mathcal{H}$ by $\Gamma((f_1, f_2)) = f_1 + f_2$.

The map Γ is a linear surjection with kernel $\mathcal{N}$ and hence, $\Gamma : \mathcal{N}^\perp \to \mathcal{H}$ is a vector space isomorphism. If we endow $\mathcal{H}$ with the norm and inner product that comes from this identification, then $\mathcal{H}$ will be a Hilbert space. If we let $P : \mathcal{H}_1 \oplus \mathcal{H}_2 \to \mathcal{N}^\perp$ denote the orthogonal projection, then for every $f = g_1 + g_2 \in \mathcal{H}$ we will have that

$$\begin{aligned}
\|f\|^2 &= \|P((g_1, g_2))\|^2_{\mathcal{H}_1 \oplus \mathcal{H}_2} \\
&= \min\{\|(g_1 + g, g_2 - g)\|^2_{\mathcal{H}_1 \oplus \mathcal{H}_2} : g \in F_0\} \\
&= \min\{\|(f_1, f_2)\|^2_{\mathcal{H}_1 \oplus \mathcal{H}_2} : f = f_1 + f_2, f_i \in \mathcal{H}_i, i = 1, 2\} \\
&= \min\{\|f_1\|^2_1 + \|f_2\|^2_2 : f = f_1 + f_2, f_i \in \mathcal{H}_i, i = 1, 2\}.
\end{aligned}$$

For any two functions, $f = f_1 + f_2, g = g_1 + g_2$ in $\mathcal{H}$ we will have that $\langle f, g \rangle_{\mathcal{H}} = \langle P((f_1, f_2)), P((g_1, g_2)) \rangle$.

It remains to see that $\mathcal{H}$ is an RKHS of functions on X with reproducing kernel K. Let $k_y^i(x) = K_i(x, y)$, so that $k_y^i \in \mathcal{H}_i$ is the kernel function. Note that if $(f, -f) \in \mathcal{N}$, then $\langle (f, -f), (k_y^1, k_y^2) \rangle = \langle f, k_y^1 \rangle_1 + \langle -f, k_y^2 \rangle_2 = f(y) - f(y) = 0$, so that $(k_y^1, k_y^2) \in \mathcal{N}^\perp$, for every $y \in X$. Thus, for any $f = f_1 + f_2 \in \mathcal{H}$, we have that $\langle f, k_y^1 + k_y^2 \rangle_{\mathcal{H}} = \langle P((f_1, f_2)), P((k_y^1, k_y^2)) \rangle = \langle P((f_1, f_2)), (k_y^1, k_y^2) \rangle = \langle f_1, k_y^1 \rangle_1 + \langle f_2, k_y^2 \rangle_2 = f_1(y) + f_2(y) = f(y)$.

Thus, $\mathcal{H}$ is an RKHS with reproducing kernel, $K(x, y) = k_y^1(x) + k_y^2(x) = K_1(x, y) + K_2(x, y)$, and the proof is complete. $\qquad\square$

Corollary 5.5. *Let $\mathcal{H}_i, i = 1, 2$ be RKHSs on X with reproducing kernels $K_i, i = 1, 2$, respectively. If $\mathcal{H}_1 \cap \mathcal{H}_2 = (0)$, then $\mathcal{H}(K_1 + K_2) = \{f_1 + f_2 : f_i \in \mathcal{H}_i\}$ with $\|f_1 + f_2\|^2 = \|f_1\|^2_1 + \|f_2\|^2_2$ is a reproducing kernel Hilbert space with kernel $K(x, y) = K_1(x, y) + K_2(x, y)$ and $\mathcal{H}_i, i = 1, 2$ are orthogonal subspaces of $\mathcal{H}$.*

5.3 Finite-dimensional RKHSs

We illustrate some applications of Aronszajn's theorem by examining finite dimensional RKHSs.

Let $\mathcal{H}$ be a finite-dimensional RKHS on X with reproducing kernel, K. If we choose an orthonormal basis for $\mathcal{H}$, $f_1, \ldots, f_n$, then by Theorem 2.4, $K(x, y) = \sum_{i=1}^{n} f_i(x)\overline{f_i(y)}$ and necessarily these functions will be linearly independent.

Conversely, let $f_i : X \to \mathbb{C}, i = 1, \ldots, n$ be linearly independent functions and set $K(x, y) = \sum_i f_i(x)\overline{f_i(y)}$. We shall use the above theorem to describe the space, $\mathcal{H}(K)$. If we let $K_i(x, y) = f_i(x)\overline{f_i(y)}$, and set $L_i(x, y) = \sum_{j \neq i} f_j(x)\overline{f_j(y)}$, then by Proposition 2.19, $\mathcal{H}(K_i) = \text{span}\{f_i\}$ and $\|f_i\|_i = 1$, where the norm is taken in $\mathcal{H}(K_i)$. Now, since $K(x, y) = K_i(x, y) + L_i(x, Y)$ and these functions are linearly independent, $\mathcal{H}(K_i) \cap \mathcal{H}(L_i) = (0)$, and by Corollary 5.5, these will be orthogonal subspaces of $\mathcal{H}(K)$. Thus, these functions are an orthonormal basis for $\mathcal{H}(K)$.

By contrast, consider the kernel $K(x, y) = f_1(x)\overline{f_1(y)} + f_2(x)\overline{f_2(y)} + (f_1(x) + f_2(x))\overline{(f_1(y) + f_2(y))}$, where f_1 and f_2 are linearly independent. By Papadakis's theorem, these functions will be a Parseval frame for $\mathcal{H}(K)$ and, hence, $\mathcal{H}(K)$ will be the 2-dimensional space spanned by f_1 and f_2. But since the three functions are not linearly independent, they cannot be an orthonormal basis. We now use Aronszajn's theorem to discover yes better the precise relationship between these functions.

Set $L_1(x, y) = f_1(x)\overline{f_1(y)} + f_2(x)\overline{f_2(y)}$ and let $L_2(x, y) = (f_1(x) + f_2(x)) \overline{(f_1(y) + f_2(y))}$. By the above reasoning, f_1, f_2 will be an orthonormal basis for $\mathcal{H}(L_1)$ and $\mathcal{H}(L_2)$ will be the span of the unit vector $f_1 + f_2$. Thus, by Aronszajn's theorem, $\|f_1\|_{\mathcal{H}(K)}^2 = min\{\|rf_1 + sf_2\|_{\mathcal{H}(L_1)}^2 + \|t(f_1 + f_2)\|_{\mathcal{H}(L_2)}^2 : f_1 = rf_1 + sf_2 + t(f_1 + f_2)\} = min\{|r|^2 + |s|^2 + |t|^2 : 1 = r + t, 0 = s + t\} = 2/3$. Similarly, $\|f_2\|_{\mathcal{H}(K)}^2 = 2/3$ and from this it follows that $\|f_1 + f_2\|_{\mathcal{H}(K)}^2 = 2/3$.

5.4 Pull-backs, restrictions and composition operators

Let X be a set, let $S \subseteq X$ be a subset and let $K : X \times X \to \mathbb{C}$ be a kernel function. Then the restriction of K to $S \times S$ is also a kernel function. Thus, we can use K to form an RKHS of functions on X or on the subset S and it is natural to ask about the relationship between these two RKHSs.

More generally, if S is any set and $\varphi : S \to X$ is a function, then we let $K \circ \varphi : S \times S \to \mathbb{C}$ denote the function given by $K \circ \varphi(s, t) = K(\varphi(s), \varphi(t))$. Formally, $K \circ \varphi$ is actually the function $K \circ (\varphi \times \varphi)$, but that notation, while logically correct, is a bit overbearing. The case $S \subseteq X$ corresponds to letting φ denote the inclusion of S, into X.

When φ is one-to-one it is easily seen that $K \circ \varphi$ is a kernel function on S. We will show below that $K \circ \varphi$ is also a kernel function on S for a general function φ. Again, it is natural to ask about the relationship between the RKHS $\mathcal{H}(K)$ of functions on X and the RKHS $\mathcal{H}(K \circ \varphi)$ of functions on S, and we will answer this question in the pull-back theorem.

Proposition 5.6. *Let* $\varphi : S \to X$ *and let* K *be a kernel function on* X. *Then* $K \circ \varphi$ *is a kernel function on* S.

Proof. Let $s_1, \ldots, s_n \in S$, let $\alpha_1, \ldots, \alpha_n \in \mathbb{C}$ and let $\{x_1, \ldots, x_p\} = \{\varphi(s_1), \ldots, \varphi(s_n)\}$, so that $p \leq n$. Set $A_k = \{i : \varphi(s_i) = x_k\}$ and let $\beta_k = \sum_{i \in A_k} \alpha_i$. Then

$$\sum_{i,j=1}^{n} \bar{\alpha}_i \alpha_j K(\varphi(s_i), \varphi(s_j)) = \sum_{k,l=1}^{p} \sum_{i \in A_k} \sum_{j \in A_l} \bar{\alpha}_i \alpha_j K(x_k, x_l)$$

$$= \sum_{k,l=1}^{p} \bar{\beta}_k \beta_l K(x_k, x_l) \geq 0.$$

Hence, $K \circ \varphi$ is a kernel function on S. $\qquad\qquad\square$

Theorem 5.7 (Pull-back theorem). *Let* X *and* S *be sets, let* $\varphi : S \to X$ *be a function and let* $K : X \times X \to \mathbb{C}$ *be a kernel function. Then* $\mathcal{H}(K \circ \varphi) = \{f \circ \varphi : f \in \mathcal{H}(K)\}$, *and for* $u \in \mathcal{H}(K \circ \varphi)$ *we have that* $\|u\|_{\mathcal{H}(K \circ \varphi)} = \min\{\|f\|_{\mathcal{H}(K)} : u = f \circ \varphi\}$.

Proof. Let $f \in \mathcal{H}(K)$, with $\|f\|_{\mathcal{H}(K)} = c$, then $f(x)\overline{f(y)} \leq c^2 K(x, y)$ in the positive definite order. Since this is an inequality of matrices over all finite sets, we see that $f \circ \varphi(s)\overline{f \circ \varphi(t)} \leq c^2 K(\varphi(s), \varphi(t))$. Hence, by 3.11, $f \circ \varphi \in \mathcal{H}(K \circ \varphi)$ with $\|f \circ \varphi\|_{\mathcal{H}(K \circ \varphi)} \leq c$.

This calculation shows that there exists a contractive, linear map, $C_\varphi : \mathcal{H}(K) \to \mathcal{H}(K \circ \varphi)$ given by $C_\varphi(f) = f \circ \varphi$.

Set $h_t(\cdot) = K(\varphi(\cdot), \varphi(t))$, so that these are the kernel functions for $\mathcal{H}(K \circ \varphi)$. Note that for any finite set of points and scalars, if $u = \sum_i \alpha_i h_{t_i}$, then $\|u\|_{\mathcal{H}(K \circ \varphi)} = \|\sum_i \alpha_i k_{\varphi(t_i)}\|_{\mathcal{H}(K)}$. It follows that there is a well-defined isometry, $\Gamma : \mathcal{H}(K \circ \varphi) \to \mathcal{H}(K)$, satisfying, $\Gamma(h_t) = k_{\varphi(t)}$.

We have that $C_\varphi \circ \Gamma$ is the identity on $\mathcal{H}(K \circ \varphi)$. Hence, for any $u \in \mathcal{H}(K \circ \varphi)$ the function $f = \Gamma(u)$ satisfies $u = f \circ \varphi$ with $\|u\| = \|f\|$ and the result follows. $\qquad\square$

Corollary 5.8 (Restriction theorem). *Let $K : X \times X \to \mathbb{C}$ be a kernel function, let $S \subseteq X$ be any nonempty subset and let $K_S : S \times S \to \mathbb{C}$ denote the restriction of K to S. Then K_S is a kernel function on S, a function belongs to $\mathcal{H}(K_S)$ if and only if it is the restriction of a function in $\mathcal{H}(K)$ to S and for $u \in \mathcal{H}(K_S)$, $\|u\|_{\mathcal{H}(K_S)} = \min\{\|f\|_{\mathcal{H}(K)} : u = f|_S\}$.*

Proof. Let $\varphi : S \to X$ be the inclusion $\varphi(s) = s$ and apply the pull-back theorem. □

Definition 5.9. Given sets X and S, a function $\varphi : S \to X$, and a kernel function $K : X \times X \to \mathbb{C}$, we call the RKHS $\mathcal{H}(K \circ \varphi)$ the $\boxed{\text{PULL-BACK}}$ of $\mathcal{H}(K)$ along φ and we call the linear map, $C_\varphi : \mathcal{H}(K) \to \mathcal{H}(K \circ \varphi)$ the $\boxed{\text{PULL-BACK MAP}}$.

We will also use $\varphi^*(\mathcal{H}(K))$ to denote the pull-back space $\mathcal{H}(K \circ \varphi)$. This notation is especially useful in cases where the kernel is not specified.

Note that

$$\langle \Gamma(h_t), k_y \rangle_{\mathcal{H}(K)} = k_{\varphi(t)}(y) = K(y, \varphi(t)) =$$
$$\overline{K(\varphi(t), y)} = \overline{k_y(\varphi(t))} = \overline{C_\varphi(k_y)(t)} = \langle h_t, C_\varphi(k_y) \rangle_{\mathcal{H}(K \circ \varphi)}.$$

Since the linear spans of such functions are dense in both Hilbert spaces, this calculation shows that $C_\varphi = \Gamma^*$. Since Γ is an isometry its range is a closed subspace and it follows that $\Gamma^* = C_\varphi$ is an isometry on the range of Γ and is 0 on the orthocomplement of the range. Such a map is called a $\boxed{\text{COISOMETRY}}$.

Thus, in the case that $S \subseteq X$ and φ is just the inclusion map, so that $K \circ \varphi = K|_S$, we see that C_φ is the coisometry that identifies the closure of the subspace of $\mathcal{H}(K)$ spanned by $\{k_y : y \in S\}$ which are functions on X, with the same set of functions regarded as functions on S.

5.4.1 Composition operators

Given sets $X_i, i = 1, 2$ and kernel functions $K_i : X_i \times X_i \to \mathbb{C}, i = 1, 2$, we wish to identify those functions $\varphi : X_1 \to X_2$ such that there is a well-defined, bounded map $C_\varphi : \mathcal{H}(K_2) \to \mathcal{H}(K_1)$ given by $C_\varphi(f) = f \circ \varphi$. Such an operator is called a $\boxed{\text{COMPOSITION OPERATOR}}$.

A characterization of the functions that define composition operators follows as a consequence case of the pull-back construction.

Theorem 5.10. *Let $X_i, i = 1, 2$ be sets, $\varphi : X_1 \to X_2$ a function and K_i : $X_i \times X_i \to \mathbb{C}, i = 1, 2$ kernel functions. Then the following are equivalent:*

(1) $\{f \circ \varphi : f \in \mathcal{H}(K_2)\} \subseteq \mathcal{H}(K_1)$;
(2) $C_\varphi : \mathcal{H}(K_2) \to \mathcal{H}(K_1)$ *is a bounded, linear operator;*
(3) *there exists a constant, $c > 0$, such that $K_2 \circ \varphi \leq c^2 K_1$.*

Moreover, in this case, $\left\| C_\varphi \right\|$ is the least such constant c.

Proof. Clearly, (2) implies (1). To see that (3) implies (2), let $f \in \mathcal{H}(K_2)$, with $\|f\| = M$. Then $f(x)\overline{f(y)} \leq M^2 K_2(x, y)$, which implies that

$$f(\varphi(x))\overline{f(\varphi(y))} \leq M^2 K_2(\varphi(x), \varphi(y)) \leq M^2 c^2 K_1(x, y).$$

Thus, it follows that $C_\varphi(f) = f \circ \varphi \in \mathcal{H}(K_1)$ with $\left\| C_\varphi(f) \right\|_1 \leq c \|f\|_2$. Hence, C_φ is bounded and $\left\| C_\varphi \right\| \leq c$.

Finally, by the pull-back theorem, (1) is equivalent to the statement that $\mathcal{H}(K_2 \circ \varphi) \subseteq \mathcal{H}(K_1)$, which is equivalent to the kernel inequality (3) by Aronszajn's inclusion theorem 5.1. The statement in that theorem regarding the norms of inclusion maps proves the last statement. $\square$

5.5 Products of kernels and tensor products of spaces

In Section 4.2 we saw that the product of two kernel functions is again a kernel function, but at that time we did not have a concrete description of the RKHS that was induced by the product. We now have enough tools at our disposal to describe the RKHS that arises from the product of kernels.

Recall that if $\mathcal{H}_i, i = 1, 2$ are Hilbert spaces, then we can form their tensor product, $\mathcal{H}_1 \otimes \mathcal{H}_2$, which is a new Hilbert space. If $\langle \cdot, \cdot \rangle_i, i = 1, 2$, denotes the respective inner products on the spaces, then to form this new space we first endow the algebraic tensor product with the inner product obtained by setting $\langle f \otimes g, h \otimes k \rangle = \langle f, h \rangle_1 \langle g, k \rangle_2$ and extending linearly and then completing the algebraic tensor product in the induced norm. One of the key facts about this completed tensor product is that it contains the algebraic tensor product faithfully; that is, the inner product satisfies $\langle u, u \rangle > 0$ for any $u \neq 0$ in the algebraic tensor product.

Now if $\mathcal{H}_1$ and $\mathcal{H}_2$ are RKHSs on sets X and S, respectively, then it is natural to want to identify an element of the algebraic tensor product, $u = \sum_{i=0}^n h_i \otimes f_i$ with the function, $\hat{u}(x, s) = \sum_{i=0}^n h_i(x) f_i(s)$. The following theorem shows that not only is this identification well-defined, but that it also extends to the completed tensor product.

Theorem 5.11. *Let $\mathcal{H}_1$ and $\mathcal{H}_2$ be RKHSs on sets X and S, with reproducing kernels K_1 and K_2. Then K given by $K((x, s), (y, t)) = K_1(x, y)K_2(s, t)$ is a kernel function on $X \times S$ and the map $u \rightarrow \hat{u}$ extends to a well-defined, linear isometry from $\mathcal{H}_1 \otimes \mathcal{H}_2$ onto the reproducing kernel Hilbert space $\mathcal{H}(K)$.*

Proof. Set $k_y^1(x) = K_1(x, y)$ and $k_t^2(s) = K_2(s, t)$. Note that if $u = \sum_{i=1}^{n} h_i \otimes f_i$, then $\langle u, k_y^1 \otimes k_t^2 \rangle_{\mathcal{H}_1 \otimes \mathcal{H}_2} = \sum_{i=1}^{n} \langle h_i, k_y^1 \rangle_{\mathcal{H}_1} \langle f_i, k_t^2 \rangle_{\mathcal{H}_2} = \hat{u}(y, t)$.

Thus, we may extend the mapping $u \rightarrow \hat{u}$ from the algebraic tensor product to the completed tensor product as follows. Given $u \in \mathcal{H}_1 \otimes \mathcal{H}_2$, define a function on $X \times S$ by setting $\hat{u}(y, t) = \langle u, k_y^1 \otimes k_t^2 \rangle_{\mathcal{H} \otimes \mathcal{H}_2}$.

It is readily seen that the set $\mathcal{L} = \{\hat{u} : u \in \mathcal{H}_1 \otimes \mathcal{H}_2\}$ is a vector space of functions on $X \times S$. Moreover, the map $u \rightarrow \hat{u}$ will be one-to-one unless there exists a nonzero $u \in \mathcal{H}_1 \otimes \mathcal{H}_2$ such that $\hat{u}(y, t) = 0$ for all $(y, t) \in X \times S$. But this latter condition would imply that u is orthogonal to the span of $\{k_y^1 \otimes k_t^2 : (y, t) \in X \times S\}$. But since the span of $\{k_y^1 : y \in X\}$ is dense in $\mathcal{H}_1$ and the span of $\{k_t^2 : t \in S\}$ is dense in $\mathcal{H}_2$, it readily follows that the span of $\{k_y^1 \otimes k_t^2 : (y, t) \in X \times S\}$ is dense in $\mathcal{H}_1 \otimes \mathcal{H}_2$. Hence, if $\hat{u} = 0$, then u is orthogonal to a dense subset and so $u = 0$.

Thus, we have that the map $u \rightarrow \hat{u}$ is one-to-one from $\mathcal{H}_1 \otimes \mathcal{H}_2$ onto $\mathcal{L}$ and we may use this identification to give $\mathcal{L}$ the structure of a Hilbert space. That is, for $u, v \in \mathcal{H}_1 \otimes \mathcal{H}_2$, we set $\langle \hat{u}, \hat{v} \rangle_{\mathcal{L}} = \langle u, v \rangle_{\mathcal{H}_1 \otimes \mathcal{H}_2}$.

Finally, since for any $(y, t) \in X \times S$, we have that $\hat{u}(y, t) = \langle \hat{u}, \widehat{k_y^1 \otimes k_t^2} \rangle$, we see that $\mathcal{L}$ is a reproducing kernel Hilbert space with kernel

$$K((x, s), (y, t)) = \langle \widehat{k_y^1 \otimes k_t^2}, \widehat{k_x^1 \otimes k_s^2} \rangle_{\mathcal{L}}$$
$$= \langle k_y^1 \otimes k_t^2, k_x^1 \otimes k_s^2 \rangle_{\mathcal{H}_1 \otimes \mathcal{H}_2} = \langle k_y^1, k_x^1 \rangle_{\mathcal{H}_1} \langle k_t^2, k_s^2 \rangle_{\mathcal{H}_2} = K_1(x, y)K_2(s, t)$$

and so K is a kernel function.

By the uniqueness of kernels, we have that $\mathcal{L} = \mathcal{H}(K)$ as sets of functions and as Hilbert spaces. Thus, the map $u \rightarrow \hat{u}$ is an isometric linear map from $\mathcal{H}_1 \otimes \mathcal{H}_2$ onto $\mathcal{H}(K)$. $\square$

Definition 5.12. We call the four-variable function

$$K((x, s), (y, t)) = K_1(x, y)K_2(s, t)$$

the $\boxed{\text{TENSOR PRODUCT OF THE KERNELS}}$ K_1 and K_2. We denote this kernel function by $K_1 \otimes K_2$.

Thus, the above theorem can be quickly summarized by the formula,

$$\mathcal{H}(K_1 \otimes K_2) = \widehat{\mathcal{H}(K_1) \otimes \mathcal{H}(K_2)}.$$

Because of the above theorem we shall often identify the tensor product of two reproducing kernel Hilbert spaces, $\mathcal{H} \otimes \mathcal{F}$, on sets X and S with the reproducing kernel space of functions $\mathcal{H}(K)$ on $X \times S$, when what is, in fact, meant is that $\widehat{\mathcal{H} \otimes \mathcal{F}} = \mathcal{H}(K)$. We give some examples of this below.

Proposition 5.13. *Let $H^2(\mathbb{D}^2)$ denote the two-variable Hardy space introduced in Section 1.4.3. Then $\widehat{H^2(\mathbb{D}) \otimes H^2(\mathbb{D})} = H^2(\mathbb{D}^2)$.*

Proof. Denoting the variables in the i-th disk by $z_i, w_i, i = 1, 2$, we have that the kernel for $H^2(\mathbb{D}) \otimes H^2(\mathbb{D})$ is given by $K_1 \otimes K_2((z_1, z_2), (w_1, w_2)) = \frac{1}{(1-\bar{w}_1 z_1)(1-\bar{w}_2 z_2)}$, which was shown in Section 2 to be the kernel for $H^2(\mathbb{D}^2)$. Since the kernels are equal the two sets of functions are equal. $\square$

Proposition 5.14. *Let $G_1 \subseteq \mathbb{C}^n$ and $G_2 \subseteq \mathbb{C}^m$ be bounded, open sets so that $G_1 \times G_2 \subseteq \mathbb{C}^{n+m}$ is also a bounded open set and let $B^2(G_1), B^2(G_2)$ and $B^2(G_1 \times G_2)$ denote the Bergman spaces of these domains, normalized so that $\|1\| = 1$. Then $\widehat{B^2(G_1) \otimes B^2(G_2)} = B^2(G_1 \times G_2)$.*

Proof. Let A_1, A_2 and A_3 denote the normalized Lebesgue measures on G_1, G_2 and $G_1 \times G_2$, respectively. Then we have that A_3 is the product of the measures A_1 and A_2.

If for $i = 1, 2$ we let $K_i(z_i, w_i)$ denote the reproducing kernels for $B^2(G_i)$, respectively, then for each $(w_1, w_2) \in G_1 \times G_2$ it is easily checked that $K_1(z_1, w_1)K(z_2, w_2) \in B^2(G_1 \times G_2)$. Since the product of two square integrable functions is integrable, for any $f \in B^2(G_1 \times G_2)$, by Fubini's theorem, we have that

$$\int_{G_1 \times G_2} f(z_1, z_2)\overline{K_1(z_1, w_1)K_2(z_2, w_2)}dA_3(z_1, z_2) =$$

$$\int_{G_1}\int_{G_2} f(z_1, z_2)\overline{K_1(z_1, w_1)K_2(z_2, w_2)}dA_2(z_2)dA_1(z_1) =$$

$$\int_{G_1} f(z_1, w_2)\overline{K_1(z_1, w_1)}dA_1(z_1) = f(w_1, w_2).$$

Thus, the tensor product of the kernels $K_1 \otimes K_2((z_1, z_2), (w_1, w_2)) = K_1(z_1, w_1)K_2(z_2, w_2)$ is the reproducing kernel for $B^2(G_1 \times G_2)$ and so the spaces are equal. $\square$

Now let X be a set and let $K_i : X \times X \to \mathbb{C}, i = 1, 2$ be kernel functions, then by the Schur product theorem 4.8 their product, $K : X \times X \to \mathbb{C}$ given by $K(x, y) = K_1(x, y)K_2(x, y)$ is a kernel function.

Our pull-back theorem allows us to give a nice characterization of the functions in the corresponding RKHS.

Definition 5.15. We call $K(x, y) = K_1(x, y)K_2(x, y)$ the $\boxed{\text{PRODUCT OF}}$ $\boxed{\text{THE KERNELS}}$ and we denote it by $K = K_1 \odot K_2$.

Given $K_i : X \times X \to \mathbb{C}, i = 1, 2$ we have two kernels, $K_1 \otimes K_2$, and $K_1 \odot K_2$, and two induced RKHSs. We have characterized the space $\mathcal{H}(K_1 \otimes K_2)$. We will now characterize the space $\mathcal{H}(K_1 \odot K_2)$.

Let $\Delta : X \to X \times X$ denote the $\boxed{\text{DIAGONAL MAP}}$, defined by $\Delta(x) = (x, x)$. Then $K_1 \odot K_2(x, y) = K_1 \otimes K_2(\Delta(x), \Delta(y))$, that is, $K_1 \odot K_2 = (K_1 \otimes K_2) \circ \Delta$. Thus, $\mathcal{H}(K_1 \odot K_2)$ is the pull-back of $\mathcal{H}(K_1 \otimes K_2) = \widehat{\mathcal{H}(K_1) \otimes \mathcal{H}(K_2)}$ along the diagonal map.

The following theorem summarizes these facts.

Theorem 5.16 (Products of kernels). *Let* $K_i : X \times X \to \mathbb{C}, i = 1, 2$ *be kernel functions and let* $K_1 \odot K_2(x, y) = K_1(x, y)K_2(x, y)$ *denote their product. Then* $f \in \mathcal{H}(K_1 \odot K_2)$ *if and only if* $f(x) = \hat{u}(x, x)$ *for some* $u \in \mathcal{H}(K_1) \otimes \mathcal{H}(K_2)$. *Moreover,*

$$\|f\|_{\mathcal{H}(K_1 \odot K_2)} = \min\{\|u\|_{\mathcal{H}(K_1) \otimes \mathcal{H}(K_2)} : f(x) = \hat{u}(x, x)\}.$$

As an application of this last corollary, we note the following fact, first observed in [8].

Corollary 5.17. $B^2(\mathbb{D}) = \{f(z, z) : f \in H^2(\mathbb{D}^2)\}$ *and given* $g \in B^2(\mathbb{D})$,

$$\|g\|_{B^2(\mathbb{D})} = \min\{\|f\|_{H^2(\mathbb{D}^2)} : g(z) = f(z, z)\}.$$

Proof. The result follows from the last result, the fact that the kernel for the Bergman space on the disk is the square of the kernel of the Hardy space, and the earlier identification of $H^2(\mathbb{D}) \otimes H^2(\mathbb{D}) = H^2(\mathbb{D}^2)$. $\qquad \square$

5.6 Push-outs of RKHSs

Given an RKHS $\mathcal{H}(K)$ on X and a function $\psi : X \to S$ we would also like to induce an RKHS on S. To carry out this construction, we first consider the subspace

$$\widetilde{\mathcal{H}} = \{f \in \mathcal{H}(K) : f(x_1) = f(x_2) \text{ whenever } \psi(x_1) = \psi(x_2)\}.$$

If $\tilde{K}(x, y)$ denotes the reproducing kernel for this subspace and we set $\tilde{k}_y(x) = \tilde{K}(x, y)$, then it readily follows that, whenever $\psi(x_1) = \psi(x_2)$ and $\psi(y_1) = \psi(y_2)$, we have that

$$\tilde{k}_y(x_1) = \tilde{k}_y(x_2) \text{ and } \tilde{k}_{y_1} = \tilde{k}_{y_2}.$$

Thus, for any such pair of points, $\tilde{K}(x_1, y_1) = \tilde{K}(x_2, y_2)$. It follows that there is a well-defined positive definite function on $K_\psi : S \times S \to \mathbb{C}$ given by $K_\psi(s, t) = \tilde{K}(\psi^{-1}(s), \psi^{-1}(t))$.

Definition 5.18. We call the RKHS, $\mathcal{H}(K_\psi)$ on S, the $\boxed{\text{PUSH-OUT}}$ of $\mathcal{H}(K)$ along ψ.

We will also use $\psi_*(\mathcal{H}(K))$ to denote the push-out space $\mathcal{H}(K_\psi)$. This notation is especially useful when the kernel function is not specified.

As an example, we consider the Bergman space on the disk $B^2(\mathbb{D})$. This space has a reproducing kernel given by the formula

$$K(z, w) = \frac{1}{(1 - \overline{w}z)^2} = \sum_{n=0}^{\infty}(n + 1)(\overline{w}z)^n.$$

If we let $\psi : \mathbb{D} \to \mathbb{D}$, be defined by $\psi(z) = z^2$, then we can pull $B^2(\mathbb{D})$ back along ψ and we can push $B^2(\mathbb{D})$ forward along ψ. We compute the kernels in each of these cases.

The kernel for the pull-back is simply

$$K \circ \psi(z, w) = \frac{1}{(1 - \overline{w}^2 z^2)^2} = \sum_{n=0}^{\infty}(n + 1)(\overline{w}z)^{2n}.$$

For the push-out, we first compute, $\widetilde{\mathcal{H}}$. Note that $\psi(z_1) = \psi(z_2)$ if and only if $z_2 = \pm z_1$. Thus, $\widetilde{\mathcal{H}} = \{f \in B^2(\mathbb{D}) : f(z) = f(-z)\}$, i.e. the subspace of even functions. This subspace is spanned by the even powers of z, and consequently, it has kernel

$$\tilde{K}(z, w) = \sum_{n=0}^{\infty}(2n + 1)(\overline{w}z)^{2n}.$$

Since, $\psi^{-1}(z) = \pm\sqrt{z}$, we see that the push-out of $B^2(\mathbb{D})$ is the new RKHS of functions on the disk with reproducing kernel

$$K_\psi(z, w) = \tilde{K}(\pm\sqrt{z}, \pm\sqrt{w}) = \sum_{n=0}^{\infty}(2n + 1)(\overline{w}z)^n.$$

Thus, the pull-back and the push-out of $B^2(\mathbb{D})$ are both spaces of analytic functions on $\mathbb{D}$ spanned by powers of z and these powers are orthogonal functions with different norms. The push-out and the pull-back are both examples of weighted Hardy spaces.

5.7 Multipliers of a RKHS

In this section we develop the theory of functions that multiply a reproducing kernel Hilbert space back into itself.

Definition 5.19. Let $\mathcal{H}_i$, $i = 1, 2$ be reproducing kernel Hilbert spaces on the same set X and let K_i, $i = 1, 2$ denote their kernel functions. A function $f : X \to \mathbb{C}$ is called a $\boxed{\text{MULTIPLIER}}$ of $\mathcal{H}_1$ into $\mathcal{H}_2$ provided that $f\mathcal{H}_1 := \{fh : h \in \mathcal{H}_1\} \subseteq \mathcal{H}_2$. We let $\mathcal{M}(\mathcal{H}_1, \mathcal{H}_2)$ denote the set of all multipliers of $\mathcal{H}_1$ into $\mathcal{H}_2$.

When $\mathcal{H}_1 = \mathcal{H}_2 = \mathcal{H}$ with reproducing kernel $K_1 = K_2 = K$, then we call a multiplier of $\mathcal{H}$ into $\mathcal{H}$, more simply, a multiplier of $\mathcal{H}$ and denote $\mathcal{M}(\mathcal{H}, \mathcal{H})$ by $\mathcal{M}(\mathcal{H})$.

Given a multiplier $f \in \mathcal{M}(\mathcal{H}_1, \mathcal{H}_2)$, we let $M_f : \mathcal{H}_1 \to \mathcal{H}_2$ denote the linear map $M_f(h) = fh$.

Clearly the set of multipliers $\mathcal{M}(\mathcal{H}_1, \mathcal{H}_2)$ is a vector space and the set of multipliers $\mathcal{M}(\mathcal{H})$ is an algebra.

Determining the multipliers of particular reproducing kernel Hilbert spaces is a deep and important area with many applications and is the basis of a great deal of research, but unfortunately the interesting examples require more knowledge of complex analysis and function theory than we have currently developed. However, the basic results characterizing multipliers are direct applications of the material that we have already developed. Consequently, we now wish to characterize which functions give rise to multipliers.

Proposition 5.20. *Let $\mathcal{H}$ be an RKHS on X with kernel K and let $f : X \to \mathbb{C}$ be a function, let $\mathcal{H}_0 = \{h \in \mathcal{H} : fh = 0\}$ and let $\mathcal{H}_1 = \mathcal{H}_0^\perp$. Set $\mathcal{H}_f = f\mathcal{H} = f\mathcal{H}_1$ and define an inner product on $\mathcal{H}_f$ by $\langle fh_1, fh_2 \rangle_f = \langle h_1, h_2 \rangle$ for $h_1, h_2 \in \mathcal{H}_1$. Then $\mathcal{H}_f$ is an RKHS on X with kernel, $K_f(x, y) = f(x)K(x, y)\overline{f(y)}$.*

Proof. By definition, $\mathcal{H}_f$ is a vector space of functions on X and the linear map, $h \to fh$ is a surjective, linear isometry from $\mathcal{H}_1$ onto $\mathcal{H}_f$. Thus, $\mathcal{H}_f$ is a Hilbert space.

Decomposing $k_y = k_y^0 + k_y^1$ with $k_y^i \in \mathcal{H}_i$, we have that $K_i(x, y) = k_y^i(x), i = 1, 2$, are the kernels for $\mathcal{H}_i, i = 1, 2$.

To see that $\mathcal{H}_f$ is an RKHS, note that for any fixed $y \in X$ and $h \in \mathcal{H}_1, f(y)h(y) = f(y)\langle h, k_y \rangle = f(y)\langle fh, fk_y \rangle_f = \langle fh, \overline{f(y)}fk_y^1 \rangle_f$, which shows that evaluation at y is a bounded linear functional in $\|\cdot\|_f$, and that $K_f(x, y) = \overline{f(y)}f(x)k_y^1(x)$. However, $fk_y^0 = 0$, and hence, $f(x)K_0(x, y)\overline{f(y)} = 0$, from which the result follows. $\qquad\square$

We are now in a position to characterize multipliers.

Theorem 5.21. *Let $\mathcal{H}_i, i = 1, 2$ be RKHSs on X with kernels $K_i, i = 1, 2$ and let $f : X \to \mathbb{C}$. The following are equivalent:*

(i) $f \in \mathcal{M}(\mathcal{H}_1, \mathcal{H}_2)$;
(ii) $f \in \mathcal{M}(\mathcal{H}_1, \mathcal{H}_2)$, and M_f is a bounded operator, i.e. $M_f \in B(\mathcal{H}_1, \mathcal{H}_2)$;
(iii) there exists a constant $c \geq 0$, such that $f(x)K_1(x, y)\overline{f(y)} \leq c^2 K_2(x, y)$.

Moreover, in these cases, $\|M_f\|$ is the least constant, c satisfying the inequality in (iii).

Proof. Clearly, (ii) implies (i).

$(i) \Rightarrow (iii)$. By the proposition above, $\mathcal{H}_f = f\mathcal{H}_1$ with the norm and inner product defined as above is an RKHS with kernel, $K_f(x, y) = f(x)K_1(x, y)\overline{f(y)}$. But since, $\mathcal{H}_f \subseteq \mathcal{H}_2$, by Theorem 5.1, there exists a constant $c > 0$ such that $f(x)K_1(x, y)\overline{f(y)} = K_f(x, y) \leq c^2 K_2(x, y)$.

$(iii) \Rightarrow (ii)$. Since the kernel K_f of the space $\mathcal{H}_f = f\mathcal{H}_1$ is given by $K_f(x, y) = f(x)K_1(x, y)\overline{f(y)} \leq c^2 K_2(x, y)$, using Theorem 5.1 again, we have that $f\mathcal{H}_1 \subseteq \mathcal{H}_2$. Now decompose $\mathcal{H}_1 = \mathcal{H}_{1,0} \oplus \mathcal{H}_{1,1}$ where $f\mathcal{H}_{1,0} = (0)$, and given any $h \in \mathcal{H}_1$, write $h = h_0 + h_1$ with $h_i \in \mathcal{H}_{1,i}, i = 0, 1$. Then $\|fh\|_{\mathcal{H}_2} = \|fh_1\|_{\mathcal{H}_2} \leq c\|fh_1\|_{\mathcal{H}_f} = c\|h_1\|_{\mathcal{H}_{1,1}} \leq c\|h\|_{\mathcal{H}_1}$. Thus, M_f is bounded and $\|M_f\| \leq c$.

Finally, if $\|M_f\| = c$, then for any $h_1 \in \mathcal{H}_{1,1}, \|fh_1\|_{\mathcal{H}_2} \leq c\|h_1\|_{\mathcal{H}_1} = c\|fh_1\|_{\mathcal{H}_f}$. Thus, applying Theorem 5.1 again, we have that $f(x)K_1(x, y)\overline{f(y)} = K_f(x, y) \leq c^2 K_2(x, y)$ and the result follows. $\qquad\square$

Corollary 5.22. *Let $\mathcal{H}_i, i = 1, 2$ be RKHSs on X with reproducing kernels $K_i(x, y) = k_y^i(x), i = 1, 2$. If $f \in \mathcal{M}(\mathcal{H}_1, \mathcal{H}_2)$, then for every $y \in X, M_f^*(k_y^2) = \overline{f(y)}k_y^1$.*

Thus, when, $K_1 = K_2$, every kernel function is an eigenvector of M_f. Moreover, if $k_y \neq 0$ for every $y \in X$, then we have that

$$\|f\|_\infty \leq \|M_f\|,$$

so that every multiplier is a bounded function on X.

Proof. For any $h \in \mathcal{H}_1$, we have that

$$\langle h, \overline{f(y)}k_y^1 \rangle_1 = f(y)h(y) = \langle M_f(h), k_y^2 \rangle_2 = \langle h, M_f^*(k_y^2) \rangle,$$

and hence, $\overline{f(y)}k_y^1 = M_f^*(k_y^2)$.

Now, if $K_1 = K_2 = K$, then $M_f^*(k_y) = \overline{f(y)}k_y$, and hence, $|f(y)| \leq \left\| M_f^* \right\| = \|M_f\|$, whenever $k_y \neq 0$ and the last inequality follows. $\qquad\square$

By the above results, we see that when $f \in \mathcal{M}(\mathcal{H})$, then for any point $y \in X$, such that $\|k_y\| \neq 0$, we can recover the values of f by the formula

$$f(y) = \frac{\langle k_y, M_f^*(k_y) \rangle}{K(y, y)} = \frac{\langle M_f(k_y), k_y \rangle}{K(y, y)}.$$

This motivates the following definition.

Definition 5.23. Let $\mathcal{H}$ be an RKHS on X with kernel $K(x, y)$, and let $T \in B(\mathcal{H})$. Then the function

$$B_T(y) = \frac{\langle T(k_y), k_y \rangle}{K(y, y)},$$

defined at any point where $K(y, y) \neq 0$, is called the $\boxed{\text{BEREZIN TRANSFORM}}$ of T.

We present one application of the Berezin transform concept

For every Hilbert space $\mathcal{H}$ there is a topology on $B(\mathcal{H})$ called the *weak operator topology*. This topology is characterized by the fact that a net $\{T_\lambda\} \subseteq B(\mathcal{H})$ converges to an operator $T \in B(\mathcal{H})$ if and only if, for every pair of vectors, $h, k \in \mathcal{H}$, $\lim_\lambda \langle T_\lambda(h), k \rangle = \langle T(h), k \rangle$. Note that if $\{T_\lambda\}$ converges in the weak topology to T, then $\{T_\lambda^*\}$ converges in the weak topology to T^*.

Corollary 5.24. *Let $\mathcal{H}$ be an RKHS on X. Then $\{M_f : f \in \mathcal{M}(\mathcal{H})\}$ is a unital subalgebra of $B(\mathcal{H})$ that is closed in the weak operator topology.*

Proof. It is easy to see that the identity operator is the multiplier corresponding to the constant function, 1, and that products and linear combinations of multipliers are multipliers. Thus, this set is a unital subalgebra of $B(\mathcal{H})$.

To see that it is closed in the weak operator topology, we must show that the limit of a net of multipliers is again a multiplier. Let $\{M_{f_\lambda}\}$ be a net of multipliers that converges in the weak operator topology to T. Then for every point y where it is defined, $\lim_\lambda f_\lambda(y) = B_T(y)$.

Set $f(y) = B_T(y)$ whenever $k_y \neq 0$, and set $f(y) = 0$ when $k_y = 0$. We claim that $T = M_f$.

To see this note that for any points $x, y \in X$,

$$(Tk_x)(y) = \langle Tk_x, k_y \rangle = \langle k_x, T^*k_y \rangle = \lim_\lambda \langle k_x, M_{f_\lambda}^* k_y \rangle$$

$$= \lim_\lambda \langle k_x, \overline{f_\lambda(y)}k_y \rangle = \langle k_x, \overline{f(y)}k_y \rangle = f(y)k_x(y).$$

Thus, T acts as multiplication by f on each kernel function and, consequently, as multiplication by f on every finite linear combination of kernel functions. If $\{h_n\}$ is a sequence of finite linear combinations of kernel functions, that converges in norm to a function h, then since T is a bounded linear operator, $(Th)(y) = \lim_n (Th_n)(y) = \lim_n f(y)h_n(y) = f(y)h(y)$. Thus, $T = M_f$ and the result follows. $\qquad\square$

5.8 Exercises

Exercise 5.1. Show that $H^2(\mathbb{D})$ is contractively contained in $B^2(\mathbb{D})$.

Exercise 5.2. Let $H_0^2(\mathbb{D}) = \{f \in H^2(\mathbb{D}) : f(0) = 0\}$. Show that $H_0^2(\mathbb{D})^\perp$ is the set of constant functions. Use this fact to compute the reproducing kernel for $H_0^2(\mathbb{D})$.

Exercise 5.3. Let $\mathcal{H}$ be an RKHS on X with reproducing kernel K, fix $x_0 \in X$ and let $\mathcal{H}_0 = \{f \in \mathcal{H} : f(x_0) = 0\}$. Compute the kernel function for $\mathcal{H}_0$.

Exercise 5.4. Let $\alpha \in \mathbb{D}$ and let $\varphi_\alpha(z) = \frac{z-\alpha}{1-\bar\alpha z}$ denote the elementary Möbius transform. Prove that the reproducing kernel for $\{f \in H^2(\mathbb{D}) : f(\alpha) = 0\}$ is $K(z, w) = \frac{\varphi_\alpha(z)\overline{\varphi_\alpha(w)}}{1-\bar{w}z}$.

Exercise 5.5. Determine $\langle f_1, f_2 \rangle_{\mathcal{H}(K)}$ for the last example. Find an orthonormal basis for this space.

Exercise 5.6. Let $X = \mathbb{D}$ and let K be the Szegö kernel. Describe the spaces $\mathcal{H}(K \circ \varphi)$ for $\varphi(z) = z^2$ and for $\varphi(z) = \frac{z-\alpha}{1-\bar\alpha z}, \alpha \in \mathbb{D}$ a simple Möbius map.

Exercise 5.7. Let $R = \{x + iy : x > 0\}$ denote the right-half plane, and let $\varphi : R \to \mathbb{D}$ be defined by $\varphi(z) = \frac{z-1}{z+1}$. Compute the pull-back of the Szegö kernel. Show that $h \in \varphi^*(H^2(\mathbb{D}))$ if and only if the function $f(z) = h(\frac{1+z}{1-z})$ is in $H^2(\mathbb{D})$.

These problems outline an alternate proof Theorem 5.21, by showing instead that $(i) \Rightarrow (ii) \Rightarrow (iii) \Rightarrow (i)$.

Exercise 5.8. Use the closed graph theorem to give a direct proof that in Theorem 5.21, (i) implies (ii).

Exercise 5.9. Note that in Theorem 5.21 M_f is bounded if and only if M_f^* is bounded, $\left\| M_f \right\| = \left\| M_f^* \right\|$ and that, in this case, $M_f^*(k_y^2) = \overline{f(y)}k_y^1$. Let $\left\| M_f^* \right\| = c$ and, for any points $x_1, \ldots, x_n \in X$ and any choice of scalars $\alpha_1, \ldots, \alpha_n$, compute $\left\| \sum_i \overline{\alpha_i} k_{x_i}^2 \right\|^2$ and use it to show that $f(x)K_1(x, y)\overline{f(y)} \le c^2 K_2(x, y)$. Thus, we have proved that (ii) implies (iii).

Exercise 5.10. Prove directly that (iii) implies (i) in Theorem 5.21 by noting that in the proof given that (iii) implies (ii), we first proved that (iii) implies (i).

Exercise 5.11. Let $\psi : \mathbb{D} \to \mathbb{D}$, be defined by $\psi(z) = z^2$, as above, and compute the pull-back and push-out of $H^2(\mathbb{D})$ along ψ.

Exercise 5.12. Let $K(x, y) = \exp(-\|x - y\|^2)$ be the Gaussian kernel on $\mathbb{R}^p$. Show that the only multipliers of $\mathcal{H}(K)$ are the constant functions.

Exercise 5.13. Give an example of a reproducing kernel Hilbert space, of dimension 2 or more, for which $\mathcal{M}(\mathcal{H}) = \mathcal{H}$ as a set. Is it possible for the norms to agree?

6

Vector-valued spaces

6.1 Basic theory

In this chapter we present some of the theory of vector-valued RKHS. The results of this chapter are not used again until we return to the study of stochastic processes, so the reader who is eager to get to the applications can safely omit this chapter and proceed directly to the chapters on applications, and then return to this chapter at a later time.

In addition to Hilbert spaces of real or complex valued functions on a given set, Hilbert spaces of vector-valued functions on a set also play an important role. Many of the theorems that we have seen in earlier chapters have straightforward generalizations to the vector-valued case. In fact, many references only present the vector-valued versions and regard the scalar versions as special cases.

However, the proofs of these vector-valued cases require some deeper results from the theory of operators on Hilbert space and tend to be much more abstract than other results that we have encountered. But because the vector-valued proofs bear so much resemblance to the proofs in earlier chapters we have chosen to present them here, while the memories of these other proofs are fresh. Also, we present them here because this chapter constitutes part of the general theory.

To do complete justice to the vector-valued theory we would need to present analogues of every theorem that we have seen for the scalar theory. That would take us too far afield and so we will content ourselves with only presenting those parts of the theory that are essential to developing an intuition or are needed for later chapters.

Definition 6.1. Let $\mathcal{C}$ be a Hilbert space and let X be a set. We denote by $\mathcal{F}(X, \mathcal{C})$ the vector space of $\mathcal{C}$-valued functions under the usual pointwise sum and scalar multiplication. A subspace $\mathcal{H} \subseteq \mathcal{F}(X, \mathcal{C})$ is called a $\mathcal{C}$-valued RKHS on X provided $\mathcal{H}$ is a Hilbert space and for every $y \in X$, the linear evaluation map $E_y : \mathcal{H} \to \mathcal{C}$ given by $E_y(f) = f(y)$ is bounded.

When we say that $\mathcal{H}$ is a vector-valued RKHS, we mean that it is a $\mathcal{C}$-valued RKHS for some Hilbert space $\mathcal{C}$. We will use the term scalar-valued RKHS when the Hilbert space $\mathcal{C} = \mathbb{C}$, i.e. the case that we have studied up until now.

We begin with the simplest example of a vector-valued RKHS.

Example 6.2. Let $\mathcal{C} = \mathbb{C}^n$ with the usual inner product and let $\mathcal{H}$ be a scalar-valued RKHS. We denote by $\mathcal{H}^n$ the direct sum of n copies of $\mathcal{H}$ with the inner product given by $\langle f, g \rangle_{\mathcal{H}^n} = \sum_{i=1}^{n} \langle f_i, g_i \rangle_{\mathcal{H}}$, where $f = (f_1, \ldots, f_n)$, and $g = (g_1, \ldots, g_n)$ are in $\mathcal{H}^n$.

We can make $\mathcal{H}^n$ a $\mathcal{C}$-valued RKHS as follows. Identify $f \in \mathcal{H}^n$ with the function from X to $\mathbb{C}^n$ given by the formula, $f(x) = (f_1(x), \ldots, f_n(x)) \in \mathbb{C}^n$. Formally, we are considering the map $L : \mathcal{H}^n \to \mathcal{F}(X, \mathbb{C}^n)$ given by this formula. Given $f, g \in \mathcal{H}^n$ we have that

$$\big(L(f+g)\big)(x) = (f_1+g_1, \ldots, f_n+g_n)(x) = f(x)+g(x) = L(f)(x)+L(g)(x),$$

so $L(f + g) = L(f) + L(g)$ and L is additive. Given $\lambda \in \mathbb{C}$,

$$\big(L(\lambda f)\big)(x) = (\lambda f1, \ldots, \lambda f_n)(x) = \lambda f(x) = \big(\lambda L(f)\big)(x),$$

so $L(\lambda f) = \lambda L(f)$ and L is a linear map. Finally, $L(f)(x) = 0$ for all x if and only if $f_1 = \cdots = f_n = 0$ so that L is one-to-one. Thus, identifying $\mathcal{H}^n$ with $L(\mathcal{H}^n)$, we have that $\mathcal{H}^n$ is a Hilbert space and a subspace of $\mathcal{F}(X, \mathbb{C}^n)$. It remains to be seen that the point evaluations are bounded.

To see this consider

$$\|f(x)\|_{\mathbb{C}^n}^2 = \sum_{i=1}^{n} |f_i(x)|^2 \leq \sum_{i=1}^{n} \|f_i\|_{\mathcal{H}}^2 \|k_x\|_{\mathcal{H}}^2$$

$$= \|k_x\|_{\mathcal{H}}^2 \sum_{i=1}^{n} \|f_i\|_{\mathcal{H}}^2$$

$$= \|k_x\|_{\mathcal{H}}^2 \|f\|_{\mathcal{H}^n}^2 .$$

Hence, $\|f(x)\|_{\mathbb{C}^n} \leq \|k_x\|_{\mathcal{H}} \|f\|_{\mathcal{H}^n}$. Hence E_x is a bounded map and $\|E_x\| \leq \|k_x\|_{\mathcal{H}}$. If we define $f \in \mathcal{H}^{(n)}$ by setting $f_1 = k_x$ and $f_i = 0$ for $i \geq 2$, then we have that $\|E_x(f)\| = k_x(x) = \|k_x\|_{\mathcal{H}}^2 = \|f\|_{\mathcal{H}^n}^2$ and so $\|E_x\| = \|f\|_{\mathcal{H}^n} = \|k_x\|_{\mathcal{H}}$.

If $\mathcal{H}$ is an RKHS of C-valued functions on some set X then for $x, y \in X$ we have that $E_x, E_y : \mathcal{H} \to C$ are bounded and so $E_x E_y^* : C \to C$ is a bounded linear map.

Definition 6.3. Let $\mathcal{H}$ be an RKHS of C-valued functions on X. We call the map $K : X \times X \to B(C)$ defined by $K(x, y) = E_x E_y^*$ the kernel function.

Our first task is to show that when $C = \mathbb{C}$ this is really our original definition of the kernel function. This exercise will, hopefully, convince the reader why we preferred to treat scalar-valued spaces separately from vector-valued spaces.

First note that $B(\mathbb{C})$ is a one-dimensional vector space and so there is some way that we must be able to identify $B(\mathbb{C}) = \mathbb{C}$. The natural way to do this is to notice that every linear map from $\mathbb{C}$ to $\mathbb{C}$ is given as multiplication by a complex number. To find the number that we are multiplying by we can simply apply it to the number 1. Thus, the "natural" vector space isomorphism between $B(\mathbb{C})$ and $\mathbb{C}$ is given by the correspondence that maps $T \in B(\mathbb{C})$ to the number $T(1) \in \mathbb{C}$. In a similar way, the space of linear maps, $B(\mathbb{C}, \mathcal{H})$, can be identified with $\mathcal{H}$, since every $T \in B(\mathbb{C}, \mathcal{H})$ is uniquely determined by the vector $T(1) \in \mathcal{H}$.

Now suppose that $\mathcal{H}$ is a scalar-valued RKHS on the set X. Given $y \in X$ the map $E_y : \mathcal{H} \to \mathbb{C}$ is a bounded linear map, recall that this is the map $E_y(f) = \langle f, k_y \rangle$. Since E_y is bounded, it has an adjoint, $E_y^* : \mathbb{C} \to \mathcal{H}$. To understand this adjoint map, let $\lambda \in \mathbb{C}$ and $f \in \mathcal{H}$, then

$$\langle f, E_y^*(\lambda) \rangle_{\mathcal{H}} = \langle E_y(f), \lambda \rangle_{\mathbb{C}} = \langle f(y), \lambda \rangle_{\mathbb{C}} = f(y)\overline{\lambda} = \overline{\lambda} \langle f, k_y \rangle_{\mathcal{H}} = \langle f, \lambda k_y \rangle_{\mathcal{H}},$$

from which we conclude that

$$E_y^*(\lambda) = \lambda k_y.$$

Thus, the map E_y^* is naturally identified with the vector $E_y^*(1) = k_y$!

Hence, the operator, $E_x E_y^* \in B(\mathbb{C})$ is multiplication by the complex number

$$E_x E_y^*(1) = E_x\big(E_y^*(1)\big) = E_x(k_y) = k_y(x) = \langle k_y, k_x \rangle = K(x, y).$$

We now turn our attention to two examples where the space is vector-valued and so the kernel functions are operator valued. These examples are perhaps the most important examples for understanding multivariate stochastic processes.

Example 6.4. Consider the earlier example of $\mathcal{H}^n$, where $\mathcal{H}$ is an RKHS with kernel K. We will compute the kernel $\widetilde{K}$ of $\mathcal{H}^n$. We identify $B(\mathbb{C}^n)$ with the

$n \times n$ matrices M_n in the usual way. Let $f = (f_1, \ldots, f_n) \in \mathcal{H}^n$ so that $E_x(f) = f(x) = (f_1(x), \ldots, f_n(x)) = (\langle f_1, k_x \rangle, \ldots, \langle f_n, k_x \rangle)$.

In order to compute $\widetilde{K}(x, y) = E_x E_y^*$ we need to compute $E_y^* : \mathbb{C}^n \to \mathcal{H}^n$. Let $v = (v_1, \ldots, v_n) \in \mathbb{C}^n$. We have

$$\left\langle f, E_y^*(v) \right\rangle_{\mathcal{H}^n} = \langle f(y), v \rangle_{\mathbb{C}^n} = \sum_{i=1}^{n} f_i(y)\overline{v_i} = \left\langle f, (v_1 k_y, \ldots, v_n k_y) \right\rangle_{\mathcal{H}^n},$$

so that $E_y^*(v) = (v_1 k_y, \ldots, v_n k_y)$.

We now compute the (i, j) entry of the matrix $\widetilde{K}(x, y) = E_x E_y^*$. Let e_i, $1 \leq i \leq n$, be the standard basis of $\mathbb{C}^n$ and compute

$$\left\langle \widetilde{K}(x, y)e_j, e_i \right\rangle_{\mathbb{C}^n} = \left\langle E_y^* e_j, E_x^* e_i \right\rangle_{\mathcal{H}^n}.$$

Note that $E_y^* e_j$ is the vector that has the function k_y in the j-th entry and is 0 elsewhere and similarly, $E_x^* e_i$ has k_x in the i-th entry and is 0 elsewhere. Thus, if $i = j$, then we get $\langle k_y, k_x \rangle = K(x, y)$, while if $i \neq j$, then we get 0. Hence, $\widetilde{K}(x, y)$ is the matrix-valued function

$$\widetilde{K}(x, y) = \begin{bmatrix} K(x, y) & 0 & \cdots & & 0 \\ 0 & \ddots & \ddots & & \vdots \\ \vdots & & \ddots & \ddots & 0 \\ 0 & \cdots & & 0 & K(x, y) \end{bmatrix} = K(x, y)I_n,$$

where I_n denotes the $n \times n$ identity matrix.

Example 6.5. This is an infinite dimensional generalization of the earlier example. Given a Hilbert space $\mathcal{H}$ we let

$$\ell^2(\mathcal{H}) := \{(f_1, f_2, \ldots) : f_n \in \mathcal{H}, \forall n \in \mathbb{N} \text{ and } \sum_{n=1}^{\infty} \|f_n\|_{\mathcal{H}}^2 < +\infty\}.$$

For $f = (f_1, f_2, \ldots)$ and $g = (g_1, g_2, \ldots)$ are in $\ell^2(\mathcal{H})$ set

$$\langle f, g \rangle_{\ell^2(\mathcal{H})} := \sum_{n=1}^{+\infty} \langle f_n, g_n \rangle_{\mathcal{H}}.$$

It can be shown that this series always converges, and defines an inner product on $\ell^2(\mathcal{H})$ that makes this space a Hilbert space. This Hilbert space can also be identified with the tensor product $\mathcal{H} \otimes \ell^2$.

Now if $\mathcal{H}$ is an RKHS of scalar-valued functions with kernel K on the set X, and for each $f = (f_1, f_2, \ldots) \in \ell^2(\mathcal{H})$ we set $f(x) = (f_1(x), f_2(x), \ldots)$, then it is easy to see that $f(x)$ is in the Hilbert space ℓ^2 of square-summable sequences. In this manner, $\ell^2(\mathcal{H})$ can be regarded as a vector-space of ℓ^2-valued functions.

Following the steps in the last example, one can easily show that with this identification $\ell^2(\mathcal{H})$ is an RKHS of ℓ^2-valued functions on X with operator-valued kernel given by

$$\widetilde{K}(x, y) = K(x, y)I_{\ell^2}.$$

We leave this verification to the exercises 6.1.

6.1.1 Matrices of operators

In the scalar-valued setting we have seen that for every set of n points the kernel function gives rise to a positive semidefinite matrix $(K(x_i, x_j))$. A similar result holds in the vector-valued case. We first introduce the notion of positivity for matrices of operators.

An operator $T \in B(\mathcal{C})$ is called positive, written $T \geq 0$, if $\langle Tx, x \rangle \geq 0$ for all $x \in \mathcal{C}$. If $T_{i,j} \in B(\mathcal{C})$ for $1 \leq i, j \leq n$, then we can form the matrix $T = (T_{i,j})$. Using the usual rules for multiplying a matrix times a vector, this matrix can be identified in a natural way with an operator on $\mathcal{C}^n$. Given

$$v = \begin{pmatrix} v_1 \\ \vdots \\ v_n \end{pmatrix} \in \mathcal{C}^n \text{ we define } T(v) = \begin{pmatrix} \sum_{j=1}^n T_{1,j} v_j \\ \vdots \\ \sum_{j=1}^n T_{n,j} v_j \end{pmatrix}. \text{ It is not hard to check}$$

that T is a bounded linear operator and that $\|T\|^2 \leq \sum_{i,j=1}^n \|T_{i,j}\|^2$. Given

another vector $w = \begin{pmatrix} w_1 \\ \vdots \\ w_n \end{pmatrix} \in \mathcal{C}^n$, we have that

$$\langle T(v), w \rangle_{\mathcal{C}^n} = \sum_{i,j=1}^n \langle T_{i,j}(v_j), w_i \rangle_{\mathcal{C}}.$$

Given a matrix of operators $T = (T_{i,j})$ we say that T is positive and write $T \geq 0$ if $\langle Tv, v \rangle_{\mathcal{C}^n} \geq 0$ for all $v \in \mathcal{C}^n$. Hence, a matrix of operators $T = (T_{i,j})$ is positive if for all choices of vectors $v_1, \ldots, v_n \in \mathcal{C}$ we have

$$\sum_{i,j=1}^n \langle T_{i,j} v_j, v_i \rangle \geq 0.$$

Proposition 6.6. *Let $\mathcal{H}$ be a $\mathcal{C}$-valued RKHS on the set X. Let K denote the associated kernel function. Then for any set of points $x_1, \ldots, x_n \in X$ we have $(K(x_i, x_j)) \geq 0$ as an operator on $\mathcal{C}^n$.*

Proof. Let $v_1, \ldots, v_n \in \mathcal{C}$ and recall that $K(x_i, x_j) = E_{x_i} E_{x_j}^*$. We have

$$\sum_{i,j=1}^{n} \langle K(x_i, x_j)v_j, v_i \rangle = \sum_{i,j=1}^{n} \langle E_{x_i} E_{x_j}^* v_j, v_i \rangle$$

$$= \sum_{i,j=1}^{n} \langle E_{x_j}^* v_j, E_{x_i}^* v_i \rangle$$

$$= \left\| \sum_{i=1}^{n} E_{x_i}^* v_i \right\|^2 \geq 0.$$

$\square$

Proposition 6.7. *Let $\mathcal{H}$ be an RKHS of $\mathcal{C}$-valued functions on X. Let $K(x, y) = E_x E_y^*$ be the kernel function, let $x_1, \ldots, x_n \in X$ and $v_1, \ldots, v_n \in \mathcal{C}$. Then the function $g : X \to \mathcal{C}$ defined by $g(x) = \sum_{i=1}^{n} K(x, x_i)v_i$ is in $\mathcal{H}$.*

Proof. Since $E_x : \mathcal{H} \to \mathcal{C}$ it follows that for every vector $v \in \mathcal{C}$ that $E_x^* v \in \mathcal{H}$. Note that

$$g(x) = \sum_{i=1}^{n} K(x, x_i)v_i = \sum_{i=1}^{n} E_x E_{x_i}^* v_i$$

$$= E_x \left(\sum_{i=1}^{n} E_{x_i}^* v_i \right)$$

Hence, $g = \sum_{i=1}^{n} E_{x_i}^* v_i \in \mathcal{H}$. $\square$

Proposition 6.8. *The functions of the form $g = \sum_{i=1}^{n} E_{x_i}^* v_i$, where $x_1, \ldots, x_n \in X$, $v_1, \ldots, v_n \in \mathcal{C}$ are dense in $\mathcal{H}$.*

Proof. Suppose that $f \in \mathcal{H}$ and that $f \perp E_x^* v$ for all $x \in X$ and $v \in \mathcal{C}$. We have

$$0 = \langle f, E_x^* v \rangle_{\mathcal{H}} = \langle f(x), v \rangle_{\mathcal{C}}.$$

Since this holds for all $v \in \mathcal{C}$, $f(x) = 0$, which holds for all $x \in X$ and hence $f = 0$. $\square$

Our next result establishes the fact that the kernel function uniquely determines the RKHS.

Proposition 6.9. *Let $\mathcal{H}, \mathcal{H}'$ be two RKHS of $\mathcal{C}$-valued functions on the set X. Let K, K' denote the corresponding kernels. If $K = K'$, then $\mathcal{H} = \mathcal{H}'$ (as sets) and $\|f\|_{\mathcal{H}} = \|f\|_{\mathcal{H}'}$.*

Proof. Since $K = K'$, we have that the function $g = \sum_{i=1}^{n} E_{x_i}^* v_i = \sum_{i=1}^{n} E_{x_i}'^* v_i$ belongs to both spaces. Such functions are dense in both $\mathcal{H}$ and in $\mathcal{H}'$. We now show that such functions have the same norm in both spaces. We have

$$\|g\|_{\mathcal{H}}^2 = \sum_{i,j=1}^{n} \left\langle E_{x_i} E_{x_j}^* v_j, v_i \right\rangle = \left\langle (K(x_i, x_j)) \begin{pmatrix} v_1 \\ \vdots \\ v_n \end{pmatrix}, \begin{pmatrix} v_1 \\ \vdots \\ v_n \end{pmatrix} \right\rangle.$$

Since $K = K'$ we see that $\|g\|_{\mathcal{H}} = \|g\|_{\mathcal{H}'}$.

Now any function in $f \in \mathcal{H}$ is a limit of a Cauchy sequence of such functions $\{g_n\}$ and this same sequence will be Cauchy in $\mathcal{H}'$. Since norm convergence implies pointwise convergence, we have that the function f is in $\mathcal{H}'$ with

$$\|f\|_{\mathcal{H}} = \lim_n \|g_n\|_{\mathcal{H}} = \lim_n \|g_n\|_{\mathcal{H}'} = \|f\|_{\mathcal{H}'}.$$

Reversing the roles of $\mathcal{H}$ and $\mathcal{H}'$ completes the proof. $\qquad\square$

Proposition 6.10. *Let $\mathcal{H}$ be a $\mathcal{C}$-valued RKHS. Then for any $x \in X$ we have $\|f(x)\|_{\mathcal{C}} \le \|f\|_{\mathcal{H}} \|K(x, x)\|^{1/2}$.*

Proof. We have

$$\|f(x)\|_{\mathcal{C}}^2 = \langle f(x), f(x) \rangle = \left\langle E_x^* E_x f, f \right\rangle \le \left\| E_x^* E_x \right\| \|f\|_{\mathcal{H}}^2 = \left\| E_x E_x^* \right\| \|f\|_{\mathcal{H}}^2.$$

Hence, $\|f(x)\|_{\mathcal{C}} \le \|K(x, x)\|^{1/2} \|f\|_{\mathcal{H}}$. $\qquad\square$

6.2 A Vector-valued version of Moore's theorem

Definition 6.11. We say that a function $K : X \times X \to B(\mathcal{C})$ is an operator-valued kernel function (or positive semidefinite) provided that for all finite sets of points $\{x_1, \dots, x_n\} \subseteq X$ we have $(K(x_i, x_j)) \ge 0$ as an operator on $\mathcal{C}^n$.

We now present a generalization of Moore's theorem to the vector-valued setting.

Theorem 6.12 (Moore's vector-valued theorem). *Let $K : X \times X \to B(\mathcal{C})$ be an operator-valued kernel function on a set X. Then there is an associated RKHS $\mathcal{H}$ of $\mathcal{C}$-valued functions on X such that K is the kernel function of $\mathcal{H}$.*

Proof. The proof is similar to the scalar-valued case. We first use the kernel function K to define a vector space of $\mathcal{C}$-valued functions together with an associated inner product.

Let W denote the set of functions of the form $g(x) = \sum_{i=1}^{n} K(x, x_i)v_i$, where $x_1, \ldots, x_n \in X$ and $v_1, \ldots, v_n \in \mathcal{C}$. We define $\phi : W \times W \to \mathbb{C}$ by

$$\phi\left(\sum_{j=1}^{n} K(\cdot, x_j)v_j, \sum_{i=1}^{n} K(\cdot, x_i)w_i\right) = \sum_{i,j=1}^{n} \langle K(x_i, x_j)v_j, w_i\rangle$$

We first need to check that ϕ is well-defined. Let $f = \sum_{j=1}^{n} K(\cdot, x_j)v_j$ be such that $f(x) = 0$ for all $x \in X$. To see that ϕ is well-defined, it is enough to show that, for any such f, we have $\phi(f, h) = \phi(h, f) = 0$ where $h = K(\cdot, y)w$. We have $\phi(f, h) = \sum_{j=1}^{n} \langle K(y, x_j)v_j, w\rangle = \langle f(y), w\rangle = 0$. Similarly $\phi(h, f) = 0$.

It is fairly straightforward to check that ϕ is linear in the first variable and conjugate-linear in the second variable.

We claim that $\phi(f, f) = 0$ if and only if $f = 0$. We have already seen one half of this implication. Now suppose that $\phi(f, f) = \sum_{i,j=1}^{n} \langle K(x_i, x_j)v_j, v_i\rangle = 0$. From Cauchy–Schwarz (applied to a semi-inner product) we see that $\phi(f, h) = 0$ for all $h \in \mathcal{H}$. Now let $h = K(\cdot, y)w$ and note that $0 = \phi(f, h) = \sum_{j=1}^{n} \langle K(y, x_j)v_j, w\rangle_{\mathcal{C}} = \langle f(y), w\rangle_{\mathcal{C}}$. Since this is true for all $w \in \mathcal{C}$ we see that $f(y) = 0$ for all $y \in X$ and hence, f is the function that is identically 0.

The remainder of the proof amounts to showing that the completion of the space W with respect to this inner product is an RKHS with kernel K. The remainder of the proof is similar to the scalar-valued version of Moore's theorem and we leave these details to the reader 6.2. $\qquad\square$

Thus, just as in the scalar case, each operator-valued kernel $K : X \times X \to B(\mathcal{C})$ uniquely determines an RKHS of $\mathcal{C}$-valued functions. We shall denote that RKHS as $\mathcal{H}(K)$. Now that we have a vector-valued version of Moore's theorem, it is possible to give many examples of vector-valued RKHS spaces simply by creating operator-valued kernel functions. Also, as with the earlier version of Moore's theorem, the reconstruction problem, i.e. determining the vector-valued functions that belong to the space, is a fundamental problem in the area.

We give some examples below.

Example 6.13. Let $\mathcal{H}$ be an RKHS of scalar-valued functions on the set X with kernel K and let $\mathcal{C}$ be any Hilbert space. Then setting

$$\widetilde{K} : X \times X \to B(\mathcal{C}) \text{ by } \widetilde{K}(x, y) = K(x, y)I_{\mathcal{C}},$$

defines an operator-valued kernel on X.

The resulting RKHS of $\mathcal{C}$-valued functions can be identified with the Hilbert space tensor product $\mathcal{H} \otimes \mathcal{C}$ in a manner outlined in the Exercises 6.3.

Example 6.14. One of the more important vector-valued RKHS for theoretical purposes is based on an analogue of the Szego kernel. Fix a Hilbert space $\mathcal{C}$, finite or infinite dimensional, and let

$$\mathcal{B} := \{T \in B(\mathcal{C}) : \|T\| < 1\}$$

denote the open unit ball of $B(\mathcal{C})$. Then the function

$$K : \mathcal{B} \times \mathcal{B} \to B(\mathcal{C}) \text{ given by } K(X, Y) = (I - XY^*)^{-1},$$

is a kernel function, and consequently gives rise to an RKHS of $\mathcal{C}$-valued functions on $\mathcal{B}$. When $\mathcal{C} = \mathbb{C}^n$, then $\mathcal{B}$ is the open unit ball in the space of $n \times n$ matrices and the corresponding RKHS is a space of analytic $\mathbb{C}^n$-valued functions on the matrix ball. This space turns out, in many ways, to be the right analogue of the Hardy space for matrix balls. The proof that K is a kernel function is not immediate and requires developing some additional techniques.

6.3 Interpolation

While developing the basic theory of scalar-valued RKHS we proved some positivity conditions for a function to be an element of an RKHS. We also established some results on Hilbert space interpolation. We will now extend these results to the vector-valued setting. While many results in the vector-valued case are similar to the scalar-valued setting, and one can often provide arguments that parallel the scalar-valued setting, when the space $\mathcal{C}$ is infinite dimensional there are several key differences. Many of these arise from the fact that in the infinite dimensional setting the ranges of operators are not automatically closed.

We begin with some basic facts about ranges of operators on infinite dimensional Hilbert spaces. Given Hilbert spaces $\mathcal{H}$ and $\mathcal{C}$ and a bounded linear operator, $T : \mathcal{H} \to \mathcal{C}$, we let $\mathcal{K}(T)$ and $\mathcal{R}(T) = T(\mathcal{H})$ denote the kernel and range of T, respectively. It is easily checked that T maps $\mathcal{K}(T)^\perp$ one-to-one onto $\mathcal{R}(T)$. Thus, if $\mathcal{R}(T)$ is a closed subspace, then by the bounded inverse theorem, T is an isomorphism between $\mathcal{K}(T)^\perp$ and $\mathcal{R}(T)$. These considerations lead to the following key fact, which we leave to the reader.

Proposition 6.15. *Let $\mathcal{H}$ and $\mathcal{C}$ be Hilbert spaces and let $T : \mathcal{H} \to \mathcal{C}$ be a bounded, linear operator. Then the following are equivalent:*

1. *$\mathcal{R}(T)$ is closed;*
2. *$\mathcal{R}(T^*T)$ is closed;*
3. *$\mathcal{R}(T^*)$ is closed;*
4. *$\mathcal{R}(TT^*)$ is closed.*

The following result illustrates one of the key differences between the finite and infinite dimensional cases.

Proposition 6.16. *Let $\mathcal{H}$ be an RKHS of $\mathcal{C}$-valued functions on X with kernel $K : X \times X \to B(\mathcal{C})$, let $F = \{x_1, \ldots, x_n\} \subseteq X$ be a finite set of distinct points, let $\mathcal{H}_F = \{g(x) = \sum_{i=1}^{n} K(x, x_j)v_j : v_j \in \mathcal{C}\}$, and let $E_F : \mathcal{H} \to \mathcal{C}^n$ be defined by $E_F(f) = (f(x_1), \ldots, f(x_n))$. Then $\mathcal{H}_F$ is a closed subspace of $\mathcal{H}$ if and only if $\mathcal{R}(E_F)$ is a closed subspace of $\mathcal{C}^n$. If $\mathcal{C}$ is a finite dimensional space, then $\mathcal{H}_F$ is closed.*

Proof. Check that $E_F^* : \mathcal{C}^n \to \mathcal{H}$ is given by

$$E_F^*((v_1, \ldots, v_n)) = \sum_{i=1}^{n} E_{x_i}^* v_i$$

so that $\mathcal{H}_F = \mathcal{R}(E_F^*)$. Thus, by the previous result, $\mathcal{H}_F$ is closed if and only if $\mathcal{R}(E_F)$ is closed.

Finally, if $\mathcal{C}$ is finite dimensional, then $\mathcal{R}(E_F)$ is closed, since every finite dimensional space is automatically closed. $\square$

Examples exist such that $\mathcal{R}(E_x) = \{f(x) : f \in \mathcal{H}\} \subseteq \mathcal{C}$ is not a closed subspace and hence, the set of functions of the form $\{K(x, x)v : v \in \mathcal{C}\}$ need not be a closed subspace of $\mathcal{H}$. One can see from the above result that this will depend on whether or not the operator $K(x, x)$ has closed range.

We do still have one property preserved from the scalar case.

Proposition 6.17. *Let $\mathcal{H}$ be an RKHS of $\mathcal{C}$-valued functions on X with kernel $K : X \times X \to B(\mathcal{C})$ and let $F = \{x_1, \ldots, x_n\} \subseteq X$ be a finite set of distinct points. Then*

$$\mathcal{H}_F^\perp = \{f \in \mathcal{H} : f(x_i) = 0, 1 \le i \le n\}.$$

Proof. We have that $0 = \langle f, K(\cdot, x_j)v_j \rangle = \langle f, E_{x_j}^* v_j \rangle = \langle f(x_j), v_j \rangle$ for every $v_j \in \mathcal{C}$. Hence, $f(x_j) = 0$ for every $1 \le j \le n$. $\square$

We now wish to prove a result about which values can be interpolated by an element of a $\mathcal{C}$-valued RKHS. In order to prove this result we will use a classical result from operator theory, known as the Douglas factorization theorem.

Theorem 6.18 (Douglas' factorization theorem). *Let* $\mathcal{X}$, $\mathcal{Y}$ *and* $\mathcal{Z}$ *be three Hilbert spaces. Suppose that* $S \in B(\mathcal{X}, \mathcal{Z})$ *and* $T \in B(\mathcal{Y}, \mathcal{Z})$. *Then the following conditions are equivalent:*

1. $\mathcal{R}(T) \subseteq \mathcal{R}(S)$;
2. *There exists* $m > 0$ *such that* $TT^* \le m^2 SS^*$;
3. *There exists an operator* $R \in B(\mathcal{X}, \mathcal{Y})$ *such that* $T = SR$.

Moreover, $\inf\{m : TT^* \le m^2 SS^*\} = \inf\{\|R\| : T = SR\}$ *and both are attained.*

Proof. We begin by proving (1) implies (3). Let $\mathcal{N} = \ker(S) \subseteq \mathcal{X}$ and note that $S : \mathcal{N}^\perp \to \mathcal{Z}$ is a one-to-one linear operator. Given $y \in \mathcal{Y}$ we have that $T(y) \in \mathcal{R}(S)$. Hence, there exists a unique $x \in \mathcal{N}^\perp$ such that $S(x) = T(y)$. Let $R(y) = x$ so that $SR(y) = S(x) = T(y)$ for all $y \in \mathcal{Y}$. This gives a well-defined map $R : \mathcal{Y} \to \mathcal{N}^\perp \subseteq \mathcal{X}$.

We now check that R is linear. Suppose that $y_1, y_2 \in \mathcal{Y}$, then there exists a unique $x_1, x_2 \in \mathcal{N}^\perp$ such that $T(y_i) = S(x_i), i = 1, 2$. Hence $T(y_1 + y_2) = S(x_1 + x_2)$ and since $x_1 + x_2$ is the unique such element of $\mathcal{N}^\perp$, we have that $R(y_1 + y_2) = x_1 + x_2 = R(y_1) + R(y_2)$. The proof that $R(\lambda y) = \lambda R(y)$ is similar.

To prove that R is bounded we use the closed graph theorem. Suppose that $(y_n, R(y_n)) \to (y, x)$ in norm. It follows that $y_n \to y$ in norm. If $x_n = R(y_n)$, then $x_n \to x$ in norm. Since $T(y_n) = S(x_n)$, using the fact that T and S are bounded, we have that

$$T(y) = \lim_n T(y_n) = \lim_n S(x_n) = S(x),$$

and it follows that $x = R(y)$. Therefore, the graph of R is closed and so R is bounded and we have that (3) holds.

Our next step is to prove that (3) implies (2). Suppose that $T = SR$ and let $m = \|R\|$. Given a vector $z \in \mathcal{Z}$ we have $\langle TT^*z, z \rangle = \|T^*z\|^2 = \|R^*S^*z\|^2 \le \|R^*\|^2 \|S^*z\|^2 = m^2 \langle SS^*z, z \rangle$. Hence $TT^* \le m^2 SS^*$.

Since in this case $m = \|R\|$, this calculation also shows that

$$\inf\{m : TT^* \le m^2 SS^*\} \le \inf\{\|R\| : T = SR\}.$$

Finally we establish the fact that (2) implies (1). Note that the fact that (3) implies (1) is trivial. So it will be enough to prove that (2) implies (3). Suppose that $TT^* \le m^2 SS^*$. From this inequality we see that for any $z \in \mathcal{Z}$, $\|T^*z\|^2 = \langle TT^*z, z \rangle \le m^2 \langle SS^*z, z \rangle = m^2 \|S^*z\|^2$.

We wish to define a map $B : \mathcal{R}(S^*) \to \mathcal{R}(T^*)$ by $B(S^*z) = T^*(z)$. To see that the map B is well-defined, we need to show that if $S^*z_1 = S^*z_2$, then

$T(z_1) = T(z_2)$. But $S(z_1) = S(z_2)$ implies that $\|S^*(z_1 - z_2)\| = 0$, which by the inequality implies that $\|T^*(z_1 - z_2)\| = 0$, and so, $T(z_1) = T(z_2)$. Now that we see that B is well-defined it is trivial that B is linear.

From the inequality we also get that $\|B(S^*z)\| = \|T^*(z)\| \leq m \|S^*(z)\|$ and so B is a bounded, linear map on this vector subspace. It follows that B can be extended to a well-defined, bounded linear map from $B : \mathcal{R}(S^*)^- \to \mathcal{Y}$. If we let $P : \mathcal{X} \to \mathcal{R}(S^*)^-$ denote the orthogonal projection, then $BP : \mathcal{X} \to \mathcal{Y}$ is a bounded linear map with $(BP)S^* = T^*$. Taking adjoints we obtain $T = S(PB^*)$.

Thus, $T = SR$ and (3) follows.

Note that the operator R that we obtained above satisfies, $\|R\| = \|BP\| \leq \|B\| \leq m$, from which it follows that

$$\inf\{\|R\| : T = SR\} \leq \inf\{m : TT^* \leq m^2 SS^*\},$$

and so both infima are equal. Finally, if $m_0 = \inf\{m : TT^* \leq m^2 SS^*\}$, then clearly $TT^* \leq m_0^2 SS^*$, from which it follows that one infimum is attained. The other infimum is also attained since by the above argument, we can construct an operator R with $\|R\| \leq m_0$. $\qquad \square$

Corollary 6.19. *Let $\mathcal{X}$ and $\mathcal{Y}$ be Hilbert spaces and let $T \in B(\mathcal{X}, \mathcal{Y})$. Then $\mathcal{R}(T) = \mathcal{R}((TT^*)^{1/2})$.*

Proof. Since $TT^* = (TT^*)^{1/2}(TT^*)^{1/2}$ by the equivalence of (1) and (2), we have that $\mathcal{R}(T) \subseteq \mathcal{R}((TT^*)^{1/2})$ and $\mathcal{R}((TT^*)^{1/2}) \subseteq \mathcal{R}(T)$. $\qquad \square$

Here is the key interpolation result that we promised earlier.

Theorem 6.20 (Vector-valued interpolation). *Let $\mathcal{H}$ be a $\mathcal{C}$-valued RKHS on X with kernel K. Let $v_1, \ldots, v_n \in \mathcal{C}$. Then there exists a function $g \in \mathcal{H}$ such that $g(x_i) = v_i$ if and only if the vector $\begin{pmatrix} v_1 \\ \vdots \\ v_n \end{pmatrix}$ is in the range of the operator $(K(x_i, x_j))^{1/2}$.*

Proof. Define the map $T : \mathcal{H} \to \mathcal{C}^n$ by $T(f) = \begin{pmatrix} f(x_1) \\ \vdots \\ f(x_n) \end{pmatrix} = \begin{pmatrix} E_{x_1}(f) \\ \vdots \\ E_{x_n}(f) \end{pmatrix}$.

By the corollary to Douglas' factorization theorem, $\mathcal{R}(T) = \mathcal{R}((TT^*)^{1/2})$. Thus, determining when $v \in \mathcal{R}(T)$ is equivalent to determining the range of the operator $(TT^*)^{1/2}$.

We have

$$\langle T(f), w \rangle = \sum_{i=1}^{n} \langle f(x_i), w_i \rangle = \left\langle f, \sum_{i=1}^{n} E_{x_i}^* w_i \right\rangle.$$

Hence, $T^*(w) = \sum_{i=1}^{n} E_{x_i}^* w_i$. It follows that $TT^*w = T(\sum_{i=1}^{n} E_{x_i}^* w_i) =$

$$\begin{pmatrix} \sum_{i=1}^{n} E_{x_1} E_{x_i}^* w_i \\ \vdots \\ \sum_{i=1}^{n} E_{x_n} E_{x_i}^* w_i \end{pmatrix} = (K(x_i, x_j))w.$$ Hence, $TT^* = (K(x_i, x_j))$ and it

follows that $\mathcal{R}(T) = \mathcal{R}((K(x_i, x_j)^{1/2})$. The proof is complete. $\qquad\square$

Recall that given a set X the set of finite subsets $F \subseteq X$ form a net with respect to inclusion.

Proposition 6.21. *Let F be a finite subset of X and let P_F denote the orthogonal projection from $\mathcal{H}$ onto the closed linear span of the functions $\{E_x^* v : x \in F, v \in \mathcal{C}\}$. Given $f \in \mathcal{H}$, let $g_F = P_F(f)$. Then g_F converges in norm to the function f.*

Proof. Given $\epsilon > 0$, there exists a finite number of points $x_1, \ldots, x_n$ and vectors $v_1, \ldots, v_n$ such that $\left\| f - \sum_{i=1}^{n} E_{x_i}^* v_i \right\| < \epsilon$. Now let $F = \{x_1, \ldots, x_n\}$ and suppose that $F \subseteq F'$. Since $\mathcal{H}_F \subseteq \mathcal{H}_{F'}$, we see that $\|g_{F'} - f\| \leq \|g_F - f\|$. On the other hand, the point g_F is the point in the closure of $\mathcal{H}_F$ that is closest to f and so $\|g_F - f\| < \epsilon$. Hence, $g_F \to f$. $\qquad\square$

Just as in the scalar case we can introduce an ordering on vector-valued kernels. We say that $K_1 \geq K_2$ if and only if the function $K_1 - K_2$ is a kernel function.

Let $\mathcal{X}$ be a Hilbert space. Given a pair of vectors $v, w \in \mathcal{X}$ we define a rank-one linear operator $R_{v,w}(x) = \langle x, w \rangle v$. This operator is sometimes denoted $v \otimes w^*$ where the w^* is intended to indicate to the reader that w is acting as a linear functional.

Given a function $f : X \to \mathcal{C}$ we define a function $f \otimes f^* : X \times X \to B(\mathcal{C})$ by setting $f \otimes f^*(x, y) = f(x) \otimes f(y)^*$.

The following result should be reminiscent of the scalar case and we leave the proof to the reader.

Proposition 6.22. *Let $f : X \to \mathcal{C}$ be a non-zero function. Then $f \otimes f^*$ is an operator-valued kernel with*

$$\mathcal{H}(f \otimes f^*) = \{\lambda f : \lambda \in \mathbb{C}\}.$$

We can now state the main theorem that characterizes the functions that belong to a vector-valued RKHS.

Theorem 6.23. *Let $\mathcal{H}$ be a $\mathcal{C}$-valued RKHS of functions on X with kernel K. Let $f : X \to \mathcal{C}$. Then the following are equivalent:*

1. *$f \in \mathcal{H}$;*
2. *there exists a constant m such that, for any choice of points $x_1, \ldots, x_n \in X$, there exists $h \in \mathcal{H}$ such that $\|h\|_{\mathcal{H}} \leq m$ and $h(x_i) = f(x_i)$ for $1 \leq i \leq n$;*
3. *there exists a constant m such that $f(x) \otimes f(y)^* \leq m^2 K(x, y)$.*

Moreover, in this case, the least such constant satisfying (2) and (3) is equal to $\|f\|$.

Proof. First we prove that (1) implies (3). Suppose that $f \in \mathcal{H}$ and that $x_1, \ldots, x_n \in X$ and $v_1, \ldots, v_n \in \mathcal{C}$. We let $g = \sum_{i=1}^{n} E_{x_i}^* v_i$. We have

$$\|g\|^2 = \sum_{i,j=1}^{n} \left\langle K(x_i, x_j)v_j, v_i \right\rangle.$$

By the Cauchy–Schwarz inequality we know that $|\langle f, g \rangle|^2 \leq \|f\|^2 \|g\|^2$. Now,

$$\langle f, g \rangle = \sum_{i=1}^{n} \left\langle f, E_{x_i}^* v_i \right\rangle = \sum_{i=1}^{n} \left\langle f(x_i), v_i \right\rangle.$$

Hence, $|\langle f, g \rangle|^2 = \sum_{i,j=1}^{n} \langle f(x_i), v_i \rangle \overline{\langle f(x_j), v_j \rangle}$. Note that $\langle f(x) \otimes f(y)v, w \rangle = \langle \langle v, f(y) \rangle f(x), w \rangle = \langle f(x), w \rangle \langle v, f(y) \rangle$. Hence, $|\langle f, g \rangle|^2 = \sum_{i,j=1}^{n} \left\langle f(x_i) \otimes f(x_j)v_j, v_i \right\rangle$.

Thus, we have

$$\sum_{i,j=1}^{n} \left\langle f(x_i) \otimes f(x_j)v_j, v_i \right\rangle \leq \|f\|^2 \sum_{i,j=1}^{n} \left\langle K(x_i, x_j)v_j, v_i \right\rangle.$$

This also shows that the smallest m satisfying (3) is less than $\|f\|$.

To show that (3) implies (2) we need to construct the function h. In order to do this we appeal to the Douglas factorization theorem. Given a fixed finite set $F = \{x_1, \ldots, x_n\}$ of distinct points, we have that $m^2 K(x_i, x_j)) \geq (f(x_i) \otimes f(x_j)^*)$. Let $T : \mathbb{C} \to \mathcal{C}^n$ be given by $T(\lambda) = \lambda \begin{pmatrix} f(x_1) \\ \vdots \\ f(x_n) \end{pmatrix}$ and note that $T^* v = \sum_{i=1}^{n} \langle v_i, f(x_i) \rangle$. Hence,

$$\left\langle TT^* \begin{pmatrix} v_1 \\ \vdots \\ v_n \end{pmatrix}, \begin{pmatrix} w_1 \\ \vdots \\ w_n \end{pmatrix} \right\rangle = \sum_{i,j=1}^{n} \left\langle \sum_{j=1}^{n} \langle v_j, f(x_j) \rangle, \sum_{i=1}^{n} \langle w_i, f(x_i) \rangle \right\rangle$$

$$= \sum_{i,j=1}^{n} \langle f(x_i), w_i \rangle \langle v_j, f(x_j) \rangle$$

$$= \sum_{i,j=1}^{n} \langle f(x_i) \otimes f(x_j)^* v_j, w_i \rangle$$

$$= \left\langle (f(x_i) \otimes f(x_j)^*) \begin{pmatrix} v_1 \\ \vdots \\ v_n \end{pmatrix}, \begin{pmatrix} w_1 \\ \vdots \\ w_n \end{pmatrix} \right\rangle.$$

Thus, we have that $TT^* \le m^2(K(x_i, x_j))$. From earlier calculation we have seen that $(K(x_i, x_j)) = (E_{x_i} E_{x_j}^*) = BB^*$, where $B = \begin{pmatrix} E_{x_1} \\ \vdots \\ E_{x_n} \end{pmatrix}$. Hence, by the Douglas factorization theorem we see that there is an operator $R : \mathbb{C} \to \mathcal{H}$ with norm at most m such that $T = BR$. Let $h = R(1)$ and note that $\|h\| = \|R(1)\| \le m$. Now, $T(1) = \begin{pmatrix} f(x_1) \\ \vdots \\ f(x_n) \end{pmatrix}$ and $BR(1) = \begin{pmatrix} h(x_1) \\ \vdots \\ h(x_n) \end{pmatrix}$. Hence, $h(x_i) = f(x_i)$ for $1 \le i \le n$ and we have that (2) holds.

Since $\|h\|$ is less than the m appearing in (3), this proof also shows that the smallest m satisfying (2) is smaller than the smallest m satisfying (3).

The final claim is that (2) implies (1). Given a finite set F of points we know that there exists a function $\tilde{h}_F$ such that $\left\| \tilde{h}_F \right\| \le m$ and such that $\tilde{h}_F(x) = f(x)$ for all $x \in F$. By projecting $\tilde{h}_F$ onto the closure of $\mathcal{H}_F^-$, the span of functions of the form $E_x^* v$, where $x \in F$ and $v \in \mathcal{C}$, we obtain a function h_F with $\|h_F\| \le m$ and $h_F \in \mathcal{H}_F^-$. Since $h_F - \tilde{h}_F \in \mathcal{H}_F^\perp$ by Proposition 6.17 we know that $h_F(x) = \tilde{h}_F(x) = f(x)$ for all $x \in F$.

The claim is that the collection of functions h_F, obtained in this fashion, forms a Cauchy net. Let $M = \sup\{\|h_F\| : F \subseteq X, F \text{ finite}\}$. Note that $M \le m$. Given $\epsilon > 0$, choose F_0 such that $\|h_{F_0}\|^2 \ge M^2 - \epsilon^2/4$. If $F \supseteq F_0$, then h_{F_0} and h_F agree on F_0 and so $h_{F_0} \perp (h_F - h_{F_0})$. Therefore,

$$M^2 \ge \|h_F\|^2 = \|h_F - h_{F_0}\|^2 + \|h_{F_0}\|^2 \ge \|h_F - h_{F_0}\|^2 + (M^2 - \epsilon^2/4)$$

Rearranging this inequality we obtain $\|h_F - h_{F_0}\| \le \epsilon/2$.

Now if $F_1, F_2 \supseteq F_0$, then

$$\left\| h_{F_1} - h_{F_2} \right\| \leq \left\| h_{F_1} - h_{F_0} \right\| + \left\| h_{F_2} - h_{F_0} \right\| \leq \epsilon.$$

Hence, h_F is a Cauchy net and we let $h \in \mathcal{H}$ denote the limit of this net. Since $h_F(x) = f(x)$ for all $x \in F$ and all $F \subseteq X$ we see that $h(x) = f(x)$ for all $x \in X$. Hence, $f = h \in \mathcal{H}$ with $\|f\| \leq m$.

Thus, $\|f\|$ is less than the smallest m satisfying (2). These inequalities show that the smallest m appearing in (2) and (3) are both equal to $\|f\|$. □

6.4 Operations on kernels

We now turn our attention to obtaining vector-valued analogues of some of Aronszajn's theorems.

Theorem 6.24. *Let $\mathcal{H}_1$ and $\mathcal{H}_2$ be two RKHSs of $\mathcal{C}$-valued function on the set X with kernels K_1 and K_2 respectively. Then the set of functions of the form $\mathcal{H} = \{f_1 + f_2 : f_i \in \mathcal{H}_i\}$ with norm given by $\|f\|_{\mathcal{H}}^2 = \inf\{\|f_1\|_{\mathcal{H}_1}^2 + \|f_2\|_{\mathcal{H}_2}^2 : f = f_1 + f_2\}$ is an RKHS of $\mathcal{C}$-valued functions on X. In addition, the kernel K of $\mathcal{H}$ is given by $K(x, y) = K_1(x, y) + K_2(x, y)$.*

Proof. Let $\Gamma : \mathcal{H}_1 \oplus \mathcal{H}_2 \rightarrow \mathcal{H}$ be given by $\Gamma(f_1, f_2) = f_1 + f_2$. Let $\mathcal{N} = \ker(\Gamma)$ and note that $\mathcal{N} = \{(f_1, f_2) : f_1 + f_2 = 0\}$. Hence, Γ is a bijective linear map from $\mathcal{N}^\perp$ onto $\mathcal{H}$. We now endow $\mathcal{H}$ with the norm induced by Γ. Let $f = f_1 + f_2 = \Gamma(f_1, f_2)$, and note that $\|\Gamma(f_1, f_2)\|^2 = \|f_1\|_{\mathcal{H}_1}^2 + \|f_2\|_{\mathcal{H}_2}^2$. Now let $h_i \in \mathcal{H}_i$ and suppose that $h_1 + h_2 = f$. Note that $(h_1 - f_1, h_2 - f_2) \in \mathcal{N}$ and is perpendicular to (f_1, f_2). Hence, $\|(h_1, h_2)\|^2 = \|(h_1 - f_1, h_2 - f_2) + (f_1, f_2)\|^2 \geq \|(f_1, f_2)\|^2 = \|f_1\|_{\mathcal{H}_1}^2 + \|f_2\|_{\mathcal{H}_2}^2$. Thus, $\|f\|^2 = \inf \|f_1\|_{\mathcal{H}_1}^2 + \|f_2\|_{\mathcal{H}_2}^2$ where the infimum is over all pairs f_1, f_2 such that $f_1 + f_2 = f$.

We now establish the fact that the kernel of $\mathcal{H}$ is $K_1 + K_2$. Let E_x^i denote the evaluation map on the space $\mathcal{H}_i$. Let $x \in X$ and $v \in \mathcal{C}$ and consider $(E_x^{1,*}v, E_x^{2,*}v) \in \mathcal{H}_1 \oplus \mathcal{H}_2$. We claim that this element of $\mathcal{H}_1 \oplus \mathcal{H}_2$ is in fact in $\mathcal{N}^\perp$. Let $f \in N$, then

$$\left\langle (f_1, f_2), (E_x^{1,*}v, E_x^{2,*}v) \right\rangle = \left\langle f_1, E_x^{1,*}v \right\rangle + \left\langle f_2, E_x^{2,*} \right\rangle$$
$$= \langle f_1(x), v \rangle + \langle f_2(x), v \rangle = \langle f_1(x) + f_2(x), v \rangle = 0.$$

Now suppose that $(f_1, f_2) \in \mathcal{N}^\perp$. We have that $\left\langle f_1 + f_2, E_x^{1,*}v + E_x^{2,*}v \right\rangle_{\mathcal{H}} = \left\langle (f_1, f_2), (E_x^{1,*}v, E_x^{2,*}v) \right\rangle = \left\langle f_1, E_x^{1,*}v \right\rangle_{\mathcal{H}_1} + \left\langle f_2, E_x^{2,*} \right\rangle_{\mathcal{H}_2} = f_1(x) + f_2(x).$

Hence, $E_x^* v = E_x^{1,*} v + E_x^{2,*} v$ for all $x \in X$ and $v \in \mathcal{C}$. It follows that $K(x, y)v = E_x(E_y^* v) = E_x(E_y^{1,*} v + E_y^{2,*} v) = K_1(x, y)v + K_2(x, y)v$. Since this is true for all $v \in \mathcal{C}$, we have that $K(x, y) = K_1(x, y) + K_2(x, y)$. $\square$

Theorem 6.25. *Let $\mathcal{H}_1$ and $\mathcal{H}_2$ be two RKHSs of $\mathcal{C}$-valued functions on X. Let the kernels of these spaces be K_1, K_2 respectively. If there exists m such that $K_1 \leq m^2 K_2$, then $\mathcal{H}_1 \subseteq \mathcal{H}_2$ (as sets) and $\|f\|_2 \leq m \|f\|_1$.*

Conversely, if $\mathcal{H}_1 \subseteq \mathcal{H}_2$, then there exists m such that $K_1 \leq m^2 K_2$, with $m = \|\iota\|$, where $\iota : \mathcal{H}_1 \to \mathcal{H}_2$ is the inclusion map.

Proof. First assume that $K_1 \leq m^2 K_2$. If $f \in \mathcal{H}_1$, then by Theorem 6.23, in the ordering on kernels we have that

$$f(x) \otimes f(y)^* \leq \|f\|_{\mathcal{H}_1}^2 K_1(x, y) \leq m^2 \|f\|_{\mathcal{H}_1}^2 K_2(x, y).$$

Applying Theorem 6.23 again yields that $f \in \mathcal{H}_2$ with $\|f\|_{\mathcal{H}_2} \leq m \|f\|_{\mathcal{H}_1}$. Hence, $\mathcal{H}_1 \subseteq \mathcal{H}_2$ and the inclusion map from $\mathcal{H}_1$ into $\mathcal{H}_2$ is bounded with norm at most m.

For the converse assume that $\mathcal{H}_1 \subseteq \mathcal{H}_2$. From the closed graph theorem it follows that the inclusion map $\iota : \mathcal{H}_1 \to \mathcal{H}_2$ is bounded. Let $f \in \mathcal{H}_1, x \in \mathcal{X}$ and $v \in \mathcal{C}$ we have

$$\left\langle f, E_x^{1,*} v \right\rangle_{\mathcal{H}_1} = \langle f(x), v \rangle_{\mathcal{C}} = \left\langle E_x^2 \circ \iota(f), v \right\rangle_{\mathcal{C}} = \left\langle f, \iota^* \circ E_x^{2,*} v \right\rangle_{\mathcal{H}_1}$$

which holds in both $\mathcal{H}_1$ and $\mathcal{H}_2$. Hence, $\iota^*(K_2(\cdot, x)v) = \iota^*(E_x^{2,*} v) = E_x^{1,*} v = K_1(\cdot, x)v$. Let $m = \|\iota\| = \|\iota^*\|$. For any collection of vectors $v_1, \ldots, v_n$ and points $x_1, \ldots, x_n$ we have

$$\left\| \sum_{j=1}^n K_1(\cdot, x_j) v_j \right\| = \left\| \iota^* \left(\sum_{j=1}^n K_2(\cdot, x_j) v_j \right) \right\| \leq m \left\| \sum_{j=1}^n K_2(\cdot, x_j) v_j \right\|.$$

Squaring both sides yields

$$\sum_{i,j=1}^n \left\langle K_1(x_i, x_j) v_j, v_i \right\rangle_{\mathcal{C}} \leq m^2 \sum_{i,j=1}^n \left\langle K_2(x_i, x_j) v_j, v_i \right\rangle_{\mathcal{C}},$$

which shows that $K_1 \leq m^2 K_2$. $\square$

6.5 Multiplier algebras

In this section we briefly cover the vector-valued theory of multipliers.

Definition 6.26. Let $\mathcal{H}_1$ and $\mathcal{H}_2$ be RKHSs of $\mathcal{C}$-valued functions on a set X. A function $F : X \to B(\mathcal{C})$ is called a multiplier of $\mathcal{H}_1$ into $\mathcal{H}_2$ if and only if $Fh \in \mathcal{H}_2$ for all $h \in \mathcal{H}_1$. The notation Fh denotes the pointwise product $F(x)h(x)$ of the operator $F(x)$ times the vector $f(x)$. We denote the set of multipliers by $\mathcal{M}(\mathcal{H}_1, \mathcal{H}_2)$.

The next result needs a purely operator theoretic lemma.

Lemma 6.27. *Let* $v, w \in \mathcal{C}$ *and let* $A, B \in B(\mathcal{C})$. *Then,*

$$A(v \otimes w^*)B = (Av) \otimes (B^*w)^*.$$

Proof. We have

$$A(v \otimes w^*)Bx = A(\langle Bx, w \rangle v) = \langle Bx, w \rangle \, Av = \langle x, B^*w \rangle \, Av$$
$$= ((Av) \otimes (B^*w)^*)x.$$

$\square$

Theorem 6.28. *Let* $\mathcal{H}_1, \mathcal{H}_2$ *be two RKHSs of* $\mathcal{C}$*-valued functions on* X. *Let the kernels of the two spaces be denoted* K_1, K_2 *and let* $F : X \to B(\mathcal{C})$. *Then the following are equivalent:*

1. $F \in \mathcal{M}(\mathcal{H}_1, \mathcal{H}_2)$;
2. *the map* $M_F(h) = Fh$ *is bounded from* $\mathcal{H}_1$ *to* $\mathcal{H}_2$;
3. *there is a constant* c *such that* $F(x)K_1(x, y)F(y)^* \le c^2 K_2(x, y)$.

In addition, the least such constant c *is the norm of* M_F.

6.6 Exercises

Exercise 6.1. Complete the details of Example 6.5.

Exercise 6.2. Complete the proof of the vector-valued version of Moore's theorem, by showing that the completion of W with the given inner product is an RKHS of $\mathcal{C}$-valued functions with kernel K.

Exercise 6.3. Let $\mathcal{H}$ be an RKHS of scalar-valued functions on a set X with kernel K and let $\mathcal{C}$ be a Hilbert space. For each vector of the form $u = \sum_{k=1}^{K} f_k \otimes v_k$ in $\mathcal{H} \otimes \mathcal{C}$ we can define a function $\tilde{u} : X \to \mathcal{C}$ by setting $\tilde{u}(x) = \sum_{k=1}^{K} f_k(x)v_k$. Prove the following:

1. this function is independent of the way that we express u as a finite sum of elementary tensors;
2. if $u \in \mathcal{H} \otimes \mathcal{C}$ is the norm limit of a sequence $\{u_n\}$ where each $\{u_n\}$ is a finite sum of elementary tensors, then for each $x \in X$, $\lim_n \tilde{u}_n(x)$ converges in norm;

3. prove that this limit is independent of the particular way that we express u as a limit of elementary tensors;

4. show that if we set $\tilde{u}(x)$ equal to this limit, then $\{\tilde{u} : u \in \mathcal{H} \otimes \mathcal{C}\}$ defines an RKHS of $\mathcal{C}$-valued functions on $\mathcal{C}$;

5. prove that the operator-valued kernel of this space is $K(x, y)I_\mathcal{C}$.

Exercise 6.4. Prove Proposition 6.22.

Exercise 6.5. Let $f_i : X \to \mathcal{C}$ be functions, $1 \leq i \leq n$. Prove that $K : X \times X \to B(\mathcal{C})$ given by $K(x, y) = \sum_{i=1}^{n} f_i(x) \otimes f_i(y)^*$ is a kernel and that $\mathcal{H}(K)$ is the span of these functions.

Exercise 6.6. Let $\mathcal{C}$ be a Hilbert space and define $K : \mathcal{C} \times \mathcal{C} \to B(\mathcal{C})$ by $K(v, w) = v \otimes w^*$. Prove that K is a kernel. Note that $f \in \mathcal{H}(K)$ implies that $f : \mathcal{C} \to \mathcal{C}$. Prove that every $f \in \mathcal{H}(K)$ is a bounded linear operator on $\mathcal{C}$. [In fact, $\mathcal{H}(K)$ is exactly the set of Hilbert–Schmidt operators on $\mathcal{C}$ and the norm is the Hilbert–Schmidt norm!]

Exercise 6.7. Prove Theorem 6.28.

Part II
Applications and examples

7

Power series on balls and pull-backs

In Chapter 4 we saw how a power series $p(t) = \sum_{n=0}^{\infty} a_n t^n$ with $a_n \geq 0$ for every n and radius of convergence r^2 induced a kernel $K_p(x, y) = p(\langle x, y \rangle)$ on the ball of radius r in any Hilbert space. In this chapter we complete the study of these spaces by showing that these spaces can be realized as pull-back spaces. This realization has the added benefit that it gives a representation and characterization of all the functions in the space $\mathcal{H}(K_p)$. The space that it is pulled back from is based on a construction from mathematical physics known as the ⎡ FOCK SPACE ⎤. Before introducing the Fock space, we will first need to discuss infinite direct sums of Hilbert spaces.

7.1 Infinite direct sums of Hilbert spaces

Given countably many Hilbert spaces $\mathcal{H}_n$, $n \in \mathbb{N}$, each with an inner product $\langle \cdot, \cdot \rangle_n$, their ⎡ INFINITE DIRECT SUM ⎤ is a Hilbert space denoted

$$\sum_{n \in \mathbb{N}} \oplus \mathcal{H}_n$$

and is defined as follows:

- a vector $h = (h_n)_{n \in \mathbb{N}}$ belongs to this space if and only if $h_n \in \mathcal{H}_n$ for all n and $\|h\|^2 := \sum_{n \in \mathbb{N}} \|h_n\|^2_{\mathcal{H}_n} < +\infty$;
- given $h = (h_n)$, $k = (k_n)$ in this space and $\lambda \in \mathbb{C}$, we set $h + k = (h_n + k_n)$, $\lambda h = (\lambda h_n)$;
- the inner product is defined by $\langle h, k \rangle = \sum_{n \in \mathbb{N}} \langle h_n, k_n \rangle_{\mathcal{H}_n}$.

It is fairly routine to show that $\sum n \in \mathbb{N} \oplus \mathcal{H}_n$ together with these operations is a Hilbert space (Exercise 7.1).

A vector $h = (h_n)_{n \in \mathbb{N}}$ is also often frequently written as $h = h_1 \oplus h_2 \oplus \dots$ or as $h = (h_1, h_2, \dots)$.

Definition 7.1. Given any Hilbert space $\mathcal{L}$ the $\boxed{\text{FOCK SPACE OVER } \mathcal{L}}$ is denoted $\mathcal{F}(\mathcal{L})$ and is the Hilbert space that one obtains by taking the countable collection of Hilbert spaces, $\mathbb{C}, \mathcal{L}, \mathcal{L} \otimes \mathcal{L}, \mathcal{L} \otimes \mathcal{L} \otimes \mathcal{L}, \ldots$, and forming their direct sum. Given a vector $x \in \mathcal{L}$ we set $x^{\otimes k} = x \otimes \cdots \otimes x \in \mathcal{L}^{\otimes k}$.

Note that even when we start with a finite dimensional space $\mathcal{L}$ the Fock space over $\mathcal{L}$ will be an infinite dimensional Hilbert space.

7.2 The pull-back construction

Now returning to our power series $p(t) = \sum_{n=0}^{\infty} a_n t^n$ with $a_n \geq 0$ for all n and radius of convergence r^2, let $\mathcal{L}$ be a Hilbert space and let $\mathbb{B}_r \subset \mathcal{L}$ denote the open ball of radius r. Recall that the kernel $K_p : \mathbb{B}_r \times \mathbb{B}_r \to \mathbb{C}$ is defined by $K_p(x, y) = p(\langle x, y \rangle)$.

We set

$$\phi_p(x) = (\sqrt{a_0}, \sqrt{a_1}\, x, \sqrt{a_2}\, x^{\otimes 2}, \sqrt{a_3}\, x^{\otimes 3}, \ldots) := \sum_{k=0}^{\infty} \oplus \sqrt{a_k}\, x^{\otimes k},$$

where as usual we set $x^{\otimes 0} = 1$. For the moment we only regard this as a formal series, but we shall soon show convergence under appropriate hypotheses.

Recall that when we have a Hilbert space $\mathcal{H}$ and we let $K_1 : \mathcal{H} \times \mathcal{H} \to \mathbb{C}$ be defined by $K_1(v, w) = \langle v, w \rangle_{\mathcal{H}}$ then by Proposition 4.13 this defines a kernel function on $\mathcal{H}$ and the RKHS that we obtain is the space of bounded linear functionals on $\mathcal{H}$.

Theorem 7.2. *Let* $p(t) = \sum_{n=0}^{\infty} a_n t^n$ *have radius of convergence* r^2 *with* $a_n \geq 0$ *for all n, let* $\mathcal{L}$ *be a Hilbert space, let* $\mathbb{B}_r$ *be the ball of radius r in* $\mathcal{L}$, *let* $K_p : \mathbb{B}_r \times \mathbb{B}_r \to \mathbb{C}$ *be defined by* $K_p(x, y) = p(\langle x, y \rangle)$ *and let* $K_1 : \mathcal{F}(\mathcal{L}) \times \mathcal{F}(\mathcal{L}) \to \mathbb{C}$ *be defined by* $K_1(v, w) = \langle v, w \rangle_{\mathcal{F}(\mathcal{L})}$.

Then $\phi_p : \mathbb{B}_r \to \mathcal{F}(\mathcal{L})$ *and* K_p *is the pull-back of* K_1 *along* ϕ_p. *Consequently,* $f : \mathbb{B}_r \to \mathbb{C}$ *is in* $\mathcal{H}(K_p)$ *if and only if there exists a vector* $w \in \mathcal{F}(\mathcal{L})$ *such that*

$$f(x) = \langle \phi_p(x), w \rangle_{\mathcal{F}(\mathcal{L})} = \sqrt{a_0}\, w_0 + \sum_{n=1}^{\infty} \sqrt{a_n} \left\langle x^{\otimes n}, w_n \right\rangle,$$

where $w = (w_0, w_1, \ldots) \in \mathcal{F}(\mathcal{L})$. *Moreover,*

$$\|f\|_{\mathcal{H}(K_p)} = \inf\{\|w\|_{\mathcal{F}(\mathcal{L})} : f(x) = \langle \phi_p(x), w \rangle\}.$$

Proof. Note that since the norm of $x^{\otimes n}$ is $\|x\|^n$, we see that

$$\|\phi_p(x)\|^2 = \sum_{n=0}^{\infty} a_n \|x\|^{2n} = p(\|x\|^2) < +\infty.$$

Thus, for $x \in \mathbb{B}_r$, we have that $\phi_p(x)$ defines a vector in $\mathcal{F}(\mathcal{L})$.

Now if we pull-back K_1 from the Fock space along ϕ_p then we obtain a kernel $K_1 \circ \phi_p$ on $\mathbb{B}_r$ satisfying

$$K_1 \circ \phi_p(x, y) = K_1(\phi_p(x), \phi_p(y)) = p(\langle x, y \rangle) = K_p(x, y).$$

The remaining conclusions of the theorem now follow by applying our pull-back theory, in particular, Theorem 5.7. $\qquad\square$

To actually find the unique vector of minimum norm such that $f(x) = \langle \phi_p(x), w \rangle$ can be quite challenging. Note that if we let $\mathcal{R}_p \subseteq \mathcal{F}(\mathcal{L})$ denote the closed linear span of $\{\phi_p(x) : x \in \mathbb{B}_r\}$ and P denote the orthogonal projection onto this subspace then $f(x) = \langle \phi_p(x), P(w) \rangle$ and it follows readily that $P(w)$ is the unique vector of minimum norm. To get a definitive characterization of the vectors in $\mathcal{R}_p$ takes us beyond the scope of what we would like to cover, but the interested reader should consult the literature on SYMMETRIC FOCK SPACES.

To illustrate the above results in a more concrete situation, let's consider the case that $\mathcal{L} = \mathbb{C}^M$ and write $x = (z_1, \ldots, z_M)$. If we again write the above vector $w \in \mathcal{F}(\mathbb{C}^M)$ in the form

$$w = w_0 \oplus w_1 \oplus w_2 \oplus \cdots,$$

then this gives a series representation of f as

$$f(x) = \sqrt{a_0}\, w_0 + \sqrt{a_1}\, \langle x, w_1 \rangle + \sqrt{a_2}\, \langle x \otimes x, w_2 \rangle + \cdots,$$

and each term $\langle x \otimes \cdots \otimes x, w_n \rangle$ is a homogeneous polynomial of total degree n in the variables $z_1, \ldots, z_M$. To see this last statement note that if we let e_i, $1 \le i \le M$ denote the canonical orthonormal basis for $\mathbb{C}^M$ so that $\langle x, e_i \rangle = z_i$, then the vectors of the form

$$e_{i_1} \otimes \cdots \otimes e_{i_n}$$

for all possible choices of $i_1, \ldots, i_n$ are an orthonormal basis for $(\mathbb{C}^M)^{\otimes n}$. Thus, each term $\langle x^{\otimes n}, w_n \rangle$ is a linear combination of inner products with these basis vectors. Moreover,

$$\langle x^{\otimes n}, e_{i_1} \otimes \cdots \otimes e_{i_n} \rangle = \langle x, e_{i_1} \rangle \cdots \langle x, e_{i_n} \rangle = z_{i_1} \cdots z_{i_n},$$

which is a homogeneous polynomial of degree n.

7.3 Polynomials and interpolation

We now wish to study interpolation in these spaces. Recall that an RKHS $\mathcal{H}(K)$ on X is called fully interpolating, if for every set of distinct points, $\{x_1, \ldots, x_n\}$ in X and any set of complex numbers $\{\lambda_1, \ldots, \lambda_n\}$ there exists a function $f \in \mathcal{H}(K)$ such that $f(x_i) = \lambda_i$ for all i. We have shown that $\mathcal{H}(K)$ is fully interpolating if and only if for every such set of distinct points the kernel functions $k_{x_1}, \ldots, k_{x_n}$ are linearly independent, which is also equivalent to the matrix $\left(K_P(x_i, x_j)\right)$ being invertible for any choice of distinct points $x_1, \ldots, x_n$.

To prove our main interpolation result, we need to first make clear what we mean by a polynomial in this more general situation.

Definition 7.3. Let $\mathcal{L}$ be a Hilbert space and let $f : \mathcal{L} \to \mathbb{C}$ be a function. We say that f is a $\boxed{\text{HOMOGENEOUS POLYNOMIAL OF DEGREE } n}$ provided that there exists $w_n \in \mathcal{L}^{\otimes n}$ such that

$$f(x) = \left\langle x^{\otimes n}, w_n \right\rangle.$$

We call f a $\boxed{\text{POLYNOMIAL OF DEGREE AT MOST } n}$ if there exists $w_0 \in \mathbb{C}$ and vectors $w_k \in \mathcal{L}^{\otimes k}$ for $1 \leq k \leq n$ such that

$$f(x) = w_0 + \sum_{k=1}^{n} \left\langle x^{\otimes k}, w_k \right\rangle,$$

that is, if f can be written as a sum of homogeneous polynomials of degree less than or equal to n. We call f a $\boxed{\text{POLYNOMIAL FUNCTION}}$ if it is a polynomial of degree n for some n. We let $\mathcal{P}_n(\mathcal{L})$ denote the set of polynomials of degree at most n on $\mathcal{L}$ and $\mathcal{P}(\mathcal{L})$ denote the set of all polynomials.

Note that for any homogeneous polynomial of degree n, $f(x) = \langle x^{\otimes n}, w_n \rangle$, we have that $f(\lambda x) = \lambda^n f(x)$. Also, when $\mathcal{L} = \mathbb{C}^M$ and we write $x = (z_1, \ldots, z_M)$ then this definition of a homogeneous polynomial reduces to the usual definition.

Note that any function that can be written as a product of n bounded linear functionals $f_1, \ldots, f_n$ is a homogeneous polynomial of degree n, since for each functional there exists $v_i \in \mathcal{L}$ such that $f_i(x) = \langle x, v_i \rangle$ and so we have

$$f_1(x) \cdots f_n(x) = \langle x, v_1 \rangle \cdots \langle x, v_n \rangle = \left\langle x^{\otimes n}, v_1 \otimes \cdots \otimes v_n \right\rangle.$$

We saw in Chapter 4 that when $\mathcal{L}$ is finite dimensional, then every homogeneous polynomial of degree n can be written as a finite linear combination of products of n bounded linear functionals. However, when $\mathcal{L}$ is infinite dimensional this will no longer be the case for $n > 1$. This is essentially a consequence of the fact that for $\mathcal{L}$ infinite dimensional there will exist vectors in $\mathcal{L}^{\otimes n}$ that cannot be expressed as finite linear combinations of elementary tensors.

However, we will show that this generalized concept of polynomial function shares many properties with our usual concept of polynomials.

Clearly, a sum of two polynomials of degree at most n on $\mathcal{L}$ is again a polynomial of degree at most n on $\mathcal{L}$ and a scalar multiple of a polynomial of degree n is again a polynomial of degree n. Thus, $\mathcal{P}_n(\mathcal{L})$ and $\mathcal{P}(\mathcal{L})$ are both vector spaces of functions on $\mathcal{L}$.

Our next result concerns products and shows that degrees of products also behave as for ordinary polynomials.

An $\boxed{\text{ALGEBRA OF FUNCTIONS ON } X}$ is a set of functions $\mathcal{A}$ on X that is a vector space and has the property that whenever $f, g \in \mathcal{A}$, then $fg \in \mathcal{A}$.

Proposition 7.4. *Let $f : \mathcal{L} \to \mathbb{C}$ be a polynomial of degree at most n and let $g : \mathcal{L} \to \mathbb{C}$ be a polynomial of degree at most m. Then $fg : \mathcal{L} \to \mathbb{C}$ is a polynomial of degree at most $n + m$. Consequently, $\mathcal{P}(\mathcal{L})$ is an algebra of functions on $\mathcal{L}$.*

Proof. To simplify notation, when $\alpha \in \mathbb{C}$ and v is a vector, we shall write $\alpha \otimes v$ for αv and given a vector $x \in \mathcal{L}$ we set $x^{\otimes 0} = 1$.

Let $f(x) = \sum_{i=0}^{n} \langle x^{\otimes i}, w_i \rangle$ and let $g(x) = \sum_{j=0}^{m} \langle x^{\otimes j}, v_j \rangle$.

Note that for each i and j we have that

$$\langle x^{\otimes i}, w_i \rangle \langle x^{\otimes j}, v_j \rangle = \langle x^{\otimes i+j}, v_i \otimes w_j \rangle.$$

Hence, if for each $0 \leq k \leq n + m$ we set

$$u_k = \sum_{i=0}^{k} v_i \otimes w_{k-i} \in \mathcal{L}^{\otimes k},$$

then

$$f(x)g(x) = \sum_{k=0}^{n+m} \langle x^{\otimes k}, u_k \rangle.$$

$\square$

The above formula for the vectors u_k is the analog of the Cauchy product formula for the coefficients of the product of two power series.

Next we look at interpolation.

Lemma 7.5. *Let $n \geq 1$ and let $x_0, \ldots, x_n \in \mathcal{L}$ be $n + 1$ distinct points. Then there exists a polynomial f of degree n, such that $f(x_0) = 1$ and $f(x_1) = \cdots = f(x_n) = 0$.*

Proof. First, for $n = 1$, since $x_0 \neq x_1$, if we set $w_1 = x_0 - x_1 \in \mathcal{L}$, then $\langle x_0, w_1 \rangle - \langle x_1, w_1 \rangle = \|w_1\|^2 \neq 0$ and so $\langle x_0, w_1 \rangle \neq \langle x_1, w_1 \rangle$. Now set $w_0 = - \langle x_1, w_1 \rangle \in \mathbb{C}$ and let

$$f(x) = w_0 + \langle x, w_1 \rangle .$$

Then f is a polynomial of degree 1 on $\mathcal{L}$ with $f(x_0) \neq 0$ and $f(x_1) = 0$. Thus, multiplying f by a suitable scalar yields the desired function.

Now given $n \geq 2$ and $x_0, \ldots, x_n$ distinct points in $\mathcal{L}$, for each $1 \leq i \leq n$ there is a degree 1 polynomial f_i such that $f_i(x_0) = 1$ and $f_i(x_i) = 0$. By the previous lemma, setting $f(x) = f_1(x) \cdots f_n(x)$ yields a polynomial of degree at most n that has the desired properties. $\square$

Proposition 7.6. *Let $\mathcal{L}$ be a Hilbert space, let $n \geq 1$, let $x_0, \ldots, x_n \in \mathcal{L}$ and let $\alpha_0, \ldots, \alpha_n \in \mathbb{C}$. Then there exists $f \in \mathcal{P}_n(\mathcal{L})$ such that for $0 \leq i \leq n$, $f(x_i) = \alpha_i$.*

Proof. For each i, $0 \leq i \leq n$ apply the above lemma to obtain a polynomial f_i of degree n such that $f_i(x_i) = 1$ and $f_i(x_j) = 0$ for $j \neq i$. Then

$$f(x) = \alpha_0 f_0(x) + \cdots + \alpha_n f_n(x)$$

is the desired function. $\square$

Now we can state the main result on interpolation for these spaces.

Theorem 7.7. *Let $\mathcal{L}$ be a Hilbert space, let $p(t) = \sum_{n=0}^{\infty} a_n t^n$ be a power series with $a_n > 0$ for all n and radius of convergence r^2, let $\mathbb{B}_r \subset \mathcal{L}$ denote the open ball of radius r, and let $K_p : \mathbb{B}_r \times \mathbb{B}_r \to \mathbb{C}$ be defined by $K_p(x, y) = p(\langle x, y \rangle)$. Then $\mathcal{H}(K_p)$ contains $\mathcal{P}(\mathcal{L})$ and is fully interpolating.*

Proof. Given any polynomial of the form $f(x) = \sum_{k=0}^{n} \langle x^{\otimes k}, w_k \rangle$ with $w_k \in \mathcal{L}^{\otimes k}$, set

$$v = \sqrt{a_0}^{-1} w_0 \oplus \cdots \oplus \sqrt{a_n}^{-1} w_n \in \mathcal{F}(\mathcal{L}),$$

then $f(x) = \langle \phi_p(x), v \rangle$ and so $f \in \mathcal{H}(K_p)$ by Theorem 7.2.

By Proposition 7.6, polynomials interpolate any given values at any set of distinct points and $\mathcal{H}(K_p)$ is fully interpolating. $\square$

Recall that by Theorem 3.6 the fact that $\mathcal{H}(K_p)$ is fully interpolating is equivalent to the kernel functions $k_{x_1}, \ldots, k_{x_n}$ being linearly independent and to the matrix $\left(K_p(x_i, x_j)\right)$ being invertible for any choice of distinct points $x_1, \ldots, x_n$.

We are now prepared to look at some particular examples.

7.3.1 The Drury-Arveson space as a pull-back

If we apply the above theorem to the case of the power series

$$p(t) = \sum_{n=0}^{\infty} t^n = \frac{1}{1-t},$$

then $\phi : \mathbb{B}_1 \to \mathcal{F}(\mathcal{L})$ is defined by

$$\phi(x) = 1 \oplus x \oplus (x \otimes x) \oplus \cdots$$

and $K_p : \mathbb{B}_1 \times \mathbb{B}_1 \to \mathbb{C}$ satisfies

$$K_p(x, y) = K_1(\phi(x), \phi(y)) = \langle \phi(x), \phi(y) \rangle_{\mathcal{F}(\mathcal{L})}.$$

Thus, we see that

$$K_p(x, y) = 1 + \sum_{n=1}^{\infty} (\langle x, y \rangle_{\mathcal{L}})^n = \frac{1}{1 - \langle x, y \rangle},$$

the $\boxed{\text{DRURY–ARVESON KERNEL}}$. The $\boxed{\text{DRURY–ARVESON SPACE}}$ is the RKHS induced by this kernel. This leads to the following result:

Theorem 7.8. *Let $\mathcal{L}$ be a Hilbert space and let $\mathbb{B}_1 \subset \mathcal{L}$ denote its unit ball. Then the Drury–Arveson space is fully interpolating and a function $f : \mathbb{B}_1 \to \mathbb{C}$ belongs to the Drury–Arveson space if and only if it can be written in the form*

$$f(x) = \langle \phi(x), w \rangle_{\mathcal{F}(\mathcal{L})},$$

for some vector $w \in \mathcal{F}(\mathcal{L})$. Moreover, the norm of f is the minimum norm of such vectors w.

Proof. Apply Theorem 7.2 to this situation, to deduce that f is in the Drury–Arveson space if and only if it can be written as $f(x) = \langle \phi(x), w \rangle$ for some w and that the norm of f is the minimum norm over all such vectors.

This space is fully interpolating by Theorem 7.7 and by the fact that the power series for $\frac{1}{1-t}$ has strictly positive coefficients. $\square$

Since the Drury–Arveson space is fully interpolating, one consequence is that for any distinct points $x_1, \ldots, x_n$ in the open unit ball of a Hilbert space the matrix

$$\left(\frac{1}{1 - \langle x_i, x_j \rangle} \right)$$

is invertible. This fact is quite tricky to see directly.

7.3.2 The Segal–Bargmann space

Given a Hilbert space $\mathcal{L}$, since $p(x) = e^x$ has a power series representation with infinite radius of convergence and positive coefficients, we have a kernel function $K : \mathcal{L} \times \mathcal{L} \to \mathbb{C}$ given by

$$K(x, y) = e^{\langle x, y \rangle}.$$

This kernel is called the $\boxed{\text{SEGAL–BARGMANN KERNEL ON } \mathcal{L}}$ and the resulting RKHS is called the $\boxed{\text{SEGAL–BARGMANN SPACE ON } \mathcal{L}}$. Define $\phi_e : \mathcal{L} \to \mathcal{F}(\mathcal{L})$ by setting

$$\phi_e(x) = \sum_{k=0}^{\infty} \oplus \frac{1}{\sqrt{k!}} x^{\otimes k}.$$

Applying our theory to this setting yields the following result:

Theorem 7.9. *Let $\mathcal{L}$ be a Hilbert space. Then the Segal–Bargmann space on $\mathcal{L}$ is fully interpolating, a function $f : \mathcal{L} \to \mathbb{C}$ belongs to the Segal–Bargmann space if and only if it has the form*

$$f(x) = \langle \phi_e(x), w \rangle$$

for some $w \in \mathcal{F}(\mathcal{L})$ and the norm of f in the Segal–Bargmann space is the minimum norm of such a vector w.

Since the Segal–Bargmann space is fully interpolating, we have that for any $x_1, \ldots, x_n \in \mathcal{L}$, the matrix

$$\left(e^{\langle x_i, x_j \rangle} \right)$$

is positive and invertible.

7.4 Exercises

Exercise 7.1. Let $\{ \mathcal{H}_n : n \in \mathbb{N} \}$ be Hilbert spaces, let $h = (h_n), k = (k_n) \in \sum_{n \in \mathbb{N}} \oplus \mathcal{H}_n$ and let $\lambda \in \mathbb{C}$.

- Prove that $\lambda h = (\lambda h_n)$ and $h + k = (h_n + k_n)$ are in $\sum_{n \in \mathbb{N}} \oplus \mathcal{H}_n$.

- Prove that the series $\langle h, k \rangle = \sum_{n \in \mathbb{N}} \langle h_n, k_n \rangle_{\mathcal{H}_n}$ converges.
- Prove that this formula defines an inner product on $\sum_{n \in \mathbb{N}} \oplus \mathcal{H}_n$.
- Prove that $\sum_{n \in \mathbb{N}} \oplus \mathcal{H}_n$ is complete in this inner product.

Exercise 7.2. This exercise concerns the Drury–Arveson space on the ball in $\mathcal{L} = \mathbb{C}^2$ with canonical orthonormal basis e_1, e_2. Note that for $x = (z_1, z_2)$ we have that $f(z_1, z_2) = z_1 z_2 = \langle x \otimes x, e_1 \otimes e_2 \rangle = \langle x \otimes x, e_2 \otimes e_1 \rangle$. Prove that $w = 1/2[e_1 \otimes e_2 + e_2 \otimes e_1]$ is the unique vector of minimal norm satisfying $z_1 z_2 = \langle x \otimes x, w \rangle$ and conclude that $\|f\| = 1/\sqrt{2}$. Find the norm and the unique vector of minimal norm for the function $g(z_1, z_2) = z_1^2 z_2$.

Exercise 7.3. Prove the assertions of Theorem 7.9.

Exercise 7.4. Prove that the Segal–Bargmann space on $\mathbb{C}^2$ is the tensor product of the Segal–Bargmann space on $\mathbb{C}$ with itself. Show that the Segal–Bargmann space on $\mathbb{C}^M$ is the tensor product of the Segal–Bargmann space on $\mathbb{C}$ with itself M times. (This result generalizes to Segal–Bargmann spaces on infinite dimensional Hilbert spaces, but requires the introduction of the theory of tensor products of infinitely many Hilbert spaces.)

Exercise 7.5. Let $p(t) = \sum_{k=0}^m a_k t^m$ be a polynomial with $a_k > 0$ for all k, let $\mathcal{L}$ be a Hilbert space and let $K_p : \mathcal{L} \times \mathcal{L} \to \mathbb{C}$ be defined as $K_p(x, y) = p(\langle x, y \rangle)$. Prove that $\mathcal{H}(K_p) = \mathcal{P}_m(\mathcal{L})$ and that for any choice of $n \le m + 1$ distinct points $x_1, \ldots, x_n$ in $\mathcal{L}$ the kernel functions $k_{x_1}, \ldots, k_{x_n}$ are linearly independent and the matrix $\big(K_p(x_i, x_j)\big)$ is invertible.

Exercise 7.6. Show that $p(x) = 1 - \ln(1 - x)$ has a power series representation with strictly positive coefficients and radius of convergence 1. Prove that if $x_1, \ldots, x_n$ are any distinct vectors in the unit ball of a Hilbert space, then the matrix $\big(1 - \ln(1 - \langle x_i, x_j \rangle)\big)$ is positive and invertible.

8
Statistics and machine learning

In this chapter we present some applications of the theory of RKHS to statistics and machine learning. These are active areas of research where kernel functions show up repeatedly. The theory of RKHS that we have developed provides a general framework in which problems in these areas are often more clearly posed.

We will not be able to discuss every application in these areas. Instead we provide some interesting motivating examples that we believe illustrate the area and the uses of RKHS methods. We begin by discussing applications of kernels to some classic least squares problems. We then use these to introduce the ideas behind what are often referred to in the literature as the "kernel trick", "feature maps" and the associated "representer theorem".

We will then formalize these ideas and illustrate the role that kernel functions play in statistical machine learning.

8.1 Regression problems and the method of least squares

One of the oldest techniques in statistics is the method of least squares. Suppose that we are given points, i.e. vectors, $\{x_1, \ldots, x_n\} \subseteq \mathbb{R}^p$. This is often thought of as the $\boxed{\text{DATA SET}}$. Given values $\lambda_i \in \mathbb{R}$ and a function $f : \mathbb{R}^p \to \mathbb{R}$ the squared error of the function on the data is given by

$$J(f) = \sum_{i=1}^{n} (f(x_i) - \lambda_i)^2.$$

This value is often called the $\boxed{\text{LOSS}}$.

Given a set of functions from $\mathbb{R}^p$ to $\mathbb{R}$, one wishes to find (if it exists) a function in the set that minimizes the loss $J(f)$.

We will look at three cases of this problem. The case when f is assumed to be linear, when f is assumed to be affine, and the case when $p = 1$ and f is assumed to be a polynomial of degree d.

In the case that our set of functions is an RKHS $\mathcal{H}(K)$ on the domain $X = \mathbb{R}^p$, we have already studied the problem of finding the minimum of $J(f)$ over all functions in $\mathcal{H}(K)$ in Chapter 3.3. We saw there that the problem reduces to a straightforward exercise in linear algebra involving the matrix $Q = (K(x_i, x_j))$. Recall that if we take the unique decomposition of the vector $\lambda = (\lambda_1, \ldots, \lambda_n)^t = Qz + (\beta_1, \ldots, \beta_n)^t$ where $(\beta_1, \ldots, \beta_n)^t$ is in the kernel of Q, and $z = (\alpha_1, \ldots, \alpha_n)^t$ is orthogonal to the kernel of Q, then the solution to our least squares problem is the function

$$f = \sum_{i=1}^{n} \alpha_i k_{x_i},$$

where k_x are the kernel functions. It is not hard to see that z is the unique vector for which $\|\lambda - Qz\|_2$ and $\|z\|_2$ are both minimal.

If we let P denote the projection onto the range of Q, then note that $P\lambda = Qz$.

We will show in each of these three cases, that there is an RKHS that plays this central role.

8.1.1 Linear functions

First consider the case that $f : X \to \mathbb{R}$ is assumed to be linear. Recalling Proposition 2.24, if we set $K(x, y) = \langle x, y \rangle$ then the RKHS that we obtain is exactly the set of linear functionals on $\mathbb{R}^p = X$.

Given $w \in \mathbb{R}^p$, let $f_w(x) = \langle x, w \rangle$. We can now find the linear functional f_w that minimizes the loss function by applying the above method to the matrix $Q = (\langle x_i, x_j \rangle)$.

Explicitly, we let $z = (\alpha_1, \ldots, \alpha_n)^t$ be the unique solution to $P\lambda = Qz$, where P is the orthogonal projection onto the range of Q and $z \perp \ker(Q)$, then our functional is

$$f_w = \sum_{i=1}^{n} \alpha_i k_{x_i} = \sum_{i=1}^{n} \alpha_i f_{x_i}.$$

Hence,

$$w = \sum_{i=1}^{n} \alpha_i x_i,$$

and f_w is the unique linear functional that minimizes the loss function. Note that w is always in the span of the "data" vectors $\{x_1, \ldots, x_n\}$.

8.1.2 Affine functions

Now suppose instead that the objective is to find the affine function $f_{w,a} : \mathbb{R}^p \to \mathbb{R}$ where $f_{w,a}(x) = \langle x, w \rangle + a$ that minimizes the loss $J(f)$.

A well-known trick allows us to replace the affine functions by linear functions, on a space of higher dimension, and thus reduce this new problem to the problem that we just solved. This elementary trick is in fact the simplest example of what is known as the $\boxed{\text{KERNEL TRICK}}$ or $\boxed{\text{KERNELIZATION}}$. But it is really another example of a pull-back.

We begin by embedding $X = \mathbb{R}^p$ into $\mathbb{R}^p \oplus \mathbb{R} = \mathbb{R}^{p+1}$ by mapping the vector x to the vector $(x, 1)$. Let $\phi : X \to \mathbb{R}^{p+1}$ be given by $\phi(x) = (x, 1) \in \mathbb{R}^{p+1}$. Note that ϕ is a non-linear embedding. If $f_{w,a} : \mathbb{R}^p \to \mathbb{R}$ is an affine function, then

$$f_{w,a}(x) = \langle (x, 1), (w, a) \rangle_{\mathbb{R}^{p+1}} = \langle \phi(x), (w, a) \rangle.$$

Conversely, if $(v, b) \in \mathbb{R}^{p+1}$, then the function $f(x) = \langle \phi(x), (v, b) \rangle$ is affine.

Hence, we see that the set of affine functions on $X = \mathbb{R}^p$ is exactly the set of linear functions on $\mathbb{R}^{p+1}$ restricted to $\phi(X) \subseteq \mathbb{R}^{p+1}$.

That is, the set of affine functions on X is the RKHS of functions on X that arises from the pull-back along $\phi : X \to Y$ of the RKHS of linear functions on $Y = \mathbb{R}^{p+1}$. Thus, this RKHS of affine functions has kernel $\langle \phi(x), \phi(y) \rangle = \langle x, y \rangle + 1$.

Thus, the problem of minimizing the loss function $\sum_{i=1}^{n} |f(x_i) - \lambda_i|^2$ over all affine functions f on $\mathbb{R}^p$ is equivalent to minimizing $\sum_{i=1}^{n} |g(\phi(x_i)) - \lambda_i|^2$ over all *linear* functions g on $\mathbb{R}^{p+1}$. We are now back to the first problem that we studied and we know that it reduces to the above linear algebra problem, but now involving the matrix $Q = (\langle \phi(x_i), \phi(x_j) \rangle) = (\langle x_i, x_j \rangle + 1)$.

Thus, if we let $z = (\alpha_1, \ldots, \alpha_n)^t$ denote the minimum norm solution of $P\lambda = Qz$, for our new matrix Q, then our affine map will be

$$f_{w,a} = \sum_{i=1}^{n} \alpha_i k_{\phi(x_i)}.$$

Thus,

$$(w, a) = \sum_{i=1}^{n} \alpha_i \phi(x_i),$$

that is,

$$w = \sum_{i=1}^{n} \alpha_i x_i \text{ and } a = \sum_{i=1}^{n} \alpha_i.$$

Note that in this case the solution (w, a) is in the linear span of the vectors $\{\phi(x_1), \ldots, \phi(x_n)\}$.

8.1.3 Polynomials

For our third example, we look at the problem of least squares approximation by a polynomial of some given degree d. For this example we start with data $\{x_1, \ldots, x_n\} \subset \mathbb{R}$ and values $\{\lambda_1, \ldots, \lambda_n\} \subset \mathbb{R}$ and a fixed integer d. The goal is then to find a polynomial p of degree at most d for which the loss

$$J(p) = \sum_{i=1}^{n} |p(x_i) - \lambda_i|^2$$

is minimized. Since any n values can be interpolated at n points by a polynomial of degree $n - 1$, when $d \geq n - 1$ the loss can be made 0 and so the interesting case is $d + 1 < n$.

To convert this into a linear least squares problem, we consider the map $\phi : \mathbb{R} \to \mathbb{R}^{d+1}$ defined by

$$\phi(x) = (1, x, \ldots, x^d).$$

If we let $w = (w_0, \ldots, w_d) \in \mathbb{R}^{d+1}$ and let $f_w : \mathbb{R}^{d+1} \to \mathbb{R}$ be the linear functional $f_w(v) = \langle v, w \rangle$, then

$$\langle \phi(x), w \rangle = f_w(\phi(x)) = w_0 + w_1 x + \cdots + w_d x^d.$$

Thus, by letting w range over $\mathbb{R}^{d+1}$ we obtain all polynomials of degree at most d.

Again, we see that the space of polynomials of degree d on $X = \mathbb{R}$ is the pull-back along $\phi : X \to Y$ of the RKHS of linear functions on $Y = \mathbb{R}^{d+1}$.

Hence, if we consider the $n \times n$ matrix $Q = (\langle \phi(x_i), \phi(x_j) \rangle)$ we obtain the polynomial p of degree at most d that minimizes $J(p)$ by setting $p(x) = f_w(\phi(x))$ with $w = \alpha_1 \phi(x_1) + \cdots + \alpha_n \phi(x_n)$ where, as before, $z = (\alpha_1, \ldots, \alpha_n)^t$ is the vector satisfying $P\lambda = Qz$ with $z \perp \ker(Q)$.

Note that in this case the entries of Q correspond to the kernel on $\mathbb{R}$ given by

$$K(x, y) = \langle \phi(x), \phi(y) \rangle = 1 + xy + \cdots + (xy)^d,$$

which, as we've seen before, yields the RKHS on $X = \mathbb{R}$ of polynomials of degree d.

The common ingredient in each of these examples was the choice of a Hilbert space $\mathcal{L}$ and the construction of a function $\phi : X \to \mathcal{L}$ such that the functions of the form

$$f_w(\phi(x)) = \langle \phi(x), w \rangle_{\mathcal{L}} \text{ for } w \in \mathcal{L}$$

is exactly the set of functions of interest, i.e. linear, affine or polynomial of degree d.

In each case, if we set $K(x, y) = \langle \phi(x), \phi(y) \rangle_{\mathcal{L}}$ then K is a kernel function on X. Since the set of linear functionals on $Y = \mathcal{L}$ is an RKHS on Y with kernel $K_Y(v, w) = \langle v, w \rangle_{\mathcal{L}}$ and $K(x, y) = K_Y(\phi(x), \phi(y))$ by our theory of pull-backs

$$\mathcal{H}(K) = \{ f_w(\phi(x)) = \langle \phi(x), w \rangle_{\mathcal{L}} : w \in \mathcal{L} \}.$$

Finally, note that in all of these problems our loss minimization problem became a least squares problem and reduced to a linear algebra problem involving the $n \times n$ matrix $Q = (\langle \phi(x_i), \phi(x_j) \rangle) = (K(x_i, x_j))$. So in each case the least squares solution was actually the function

$$f \in \text{span}\{ k_{x_i} : 1 \le i \le n \} \subseteq \mathcal{H}(K)$$

that minimized $J(f)$ and was obtained by applying 3.3.

The easiest case of solving a loss function problem $J(f) = \sum_{i=1}^{n} |f(x_i) - \lambda_i|^2$ is when the matrix $Q = (K(x_i, x_j))$ is invertible, since in this case there will be a function f that exactly interpolates the data, i.e. the loss will be 0, and the function f is attained by setting $z = Q^{-1}\lambda$.

Suppose that our data set consists of n distinct points $x_1, \ldots, x_n$ in a Hilbert space $\mathcal{L}$. In Exercise 7.5 we saw that as long as $n \le m + 1$ then any kernel $K_p(x, y) = p(\langle x, y \rangle)$ obtained from a polynomial or power series of the form $p(t) = \sum_{k=0}^{m} a_k t^k$ in which all the coefficients $a_k > 0$ have the property that the matrix $(K_p(x_i, x_j))$ is invertible. From the results of Theorem 3.6 this invertibility is equivalent to showing that the kernel functions $k_{x_1}, \ldots, k_{x_n}$ are linearly independent.

Thus, by varying p we can design many kernels for which we will have a function f that exactly interpolates our data and for which the computation of f is a very tractable problem in linear algebra. Inverting strictly positive matrices, even for large n, is one of the best understood problems in numerical linear algebra.

8.2 The kernel method

The term $\boxed{\text{KERNEL METHOD}}$ is used as a blanket term to describe many of the techniques that we saw in the last section. We describe three aspects of the theory that are closely tied to the ideas we have developed so far.

- Feature maps. A FEATURE MAP ϕ is an embedding of the set X where the data resides into a possibly infinite-dimensional auxiliary Hilbert space $\mathcal{L}$.
- Prediction. In a PREDICTION PROBLEM the objective is to produce a function $f : X \to \mathbb{R}$ that behaves reasonably on the existing data and also makes good predictions on new data.
- Optimization. Several prediction and classification problems reduce to minimizing a risk or loss functional defined over some RKHS on X.

Given data from a set X, let $\phi : X \to \mathcal{L}$, where $\mathcal{L}$ is a Hilbert space, be a feature map. The feature map induces a kernel on X by $K(x, y) = \langle \phi(x), \phi(y) \rangle$. By choosing ϕ and $\mathcal{L}$ we are generating an RKHS $\mathcal{H}(K)$ of functions on X that become our PREDICTORS.

The advantage of this viewpoint, as we saw in the previous section, is that it can linearize some problems and often turns the problem of optimization over the possibly infinite-dimensional space of predictors, $\mathcal{H}(K)$, into a finite-dimensional linear algebra problem involving the matrix $(K(x_i, x_j))$.

Exactly how simple the problem can be made depends considerably on how complicated the loss function is. So far we have only considered loss functions that arise from least squares. However, we will see that under mild assumptions on the loss functional the optimal solution to these problems lies in a finite-dimensional subspace and is determined by the value of the kernel K on the data. This theorem is known as the REPRESENTER THEOREM 8.7 and we will prove it later in this chapter.

Thus, these kernel methods provide added flexibility in many statistical models, by allowing a large class of functions as possible predictors.

The representer theorem allows us to work with the kernel function directly and does not require the explicit computation of the feature map, which could require an infinite amount of computation for an infinite-dimensional embedding.

We will look at a simple problem in SHAPE RECOGNITION and the MAXIMAL MARGIN CLASSIFIER to illustrate this phenomenon.

We now explain a few of these ideas more fully.

Many statistical methods are concerned with making predictions from existing data. The model for data is a collection of points $x_1, \ldots, x_n$ from a set X, usually $\mathbb{R}^m$. This data is usually the result of either observations or experiments. A collection of points $x_1, \ldots, x_n$ is called a *data set*.

Associated with each of the points is a value λ_i, where $\lambda_i \in Y$. Again, it is often the case that $Y \subset \mathbb{R}$, but sometimes Y might only have two points. The prediction problem is to construct a function $f : X \to Y$ which behaves reasonably on the existing data set and will make reasonable predictions for

y on as yet unseen values of x. This last sentence is admittedly rather vague. The way that it is made precise is by introducing the loss function, which determines our measure of what "reasonable" on the data set means, and the introduction of the feature map, which determines the space of possible predictors.

A REGRESSION PROBLEM is one in which Y is an interval. When the set Y of possible values is finite $Y = \{y_1, \ldots, y_m\}$ we call this a CLASSIFICATION problem. The set of values in Y are the CLASSES . The general idea is that when the predictor function f assigns value y_i to a new point x that was not part of the original data set, then we are "predicting" that the point x belongs to the i-th class.

We will focus on a classification problem where there are only two classes.

8.3 Classification problems and geometric separation

We now look at SHAPE RECOGNITION , which is a toy example of a classical problem in machine learning. Through this example we will introduce the idea of LINEAR SEPARATION .

Let $S \subseteq \mathbb{R}^m$. We should think of S "the shape" as being fixed, but unknown to us. Starting with a finite set of points that are divided into two sets, those that we are told belong to S and those that we are told are not in S, our goal is to provide a reasonable guess as to the "shape" of the set S. To make this problem better defined, one might assume that the boundary of S is the graph of a function g that belongs to a particular set of functions. These functions are then our predictors.

To keep things concrete we will assume that the set $S \subseteq \mathbb{R}^2$ and that we know that

$$S = \{x = (\alpha, \beta) \in \mathbb{R}^2 : g(x) < 0\}$$

for some unknown function g of the form $g(x) = a + b\alpha + c\beta + d\alpha^2 + e\beta^2$. Thus, the set of functions of this form for arbitrary constants a, b, c, d, e is our set of predictors and we would like to find an optimal such function g representing our "best guess of the shape S".

Now consider the feature map $\phi : \mathbb{R}^2 \to \mathbb{R}^5$ given by

$$\phi(x) = (1, \alpha, \beta, \alpha^2, \beta^2).$$

Note that if $v = (a, b, c, d, e)$ then our predictor functions f can all be written as $f(x) = \langle \phi(x), v \rangle$ for some $v \in \mathbb{R}^5$.

This is similar to the problem that we encountered in our earlier discussion of least squares problems. By using an embedding we have linearized the problem and created an RKHS on $X = \mathbb{R}^2$ that is exactly our family of predictors: namely, the RKHS with kernel $K(x, y) = \langle \phi(x), \phi(y) \rangle$.

Note that our family of predictor functions contains all circles and ellipses in $\mathbb{R}^2$ and also many more functions. If our problem was to predict the circle that best resembles S, then our vector v would need to satisfy some additional constraints. The linear functions that correspond to circles are given by vectors v such that $v_4 = v_5$, $v_2^2 + v_3^2 - 4v_1^2 > 0$, which is not even a convex subset of $\mathbb{R}^5$. For this reason, trying to find the optimal circle that "fits" our data adds an extra level of complexity that obscures the points we wish to make.

In order to gain information about the set S we are allowed to choose a finite set of points from $\mathbb{R}^2$ and then we can ask an *oracle* whether or not the point is in the set S. Suppose that $x_1, \ldots, x_n$ is a data set. Each point in the data set is a point in $\mathbb{R}^2$. The labels λ_i are chosen as follows: if the point x_i is inside S then $\lambda_i = -1$, otherwise $\lambda_i = 1$. Note that if g is the function that determines S, then $\lambda_i = \operatorname{sign}(g(x))$.

So let $\mathcal{X}_-$ denote the points in our data set inside S and let $\mathcal{X}_+$ denote the set of points outside S.

We have available to us the data and the labels λ_i, but we do not yet know the function g, since we do not know the underlying shape S. Our problem is to find a function $f \in \mathcal{H}(K)$ such that $\operatorname{sign}(f)$ is positive at the points x_i such that $\lambda_i = 1$ and negative at the remaining points.

Since every function $f \in \mathcal{H}$ is of the form $f(x) = \langle \phi(x), v \rangle$, this is equivalent to finding a vector $v \in \mathbb{R}^5$ such that $\langle \phi(x_i), v \rangle > 0$ on the set $\mathcal{X}_+$ and $\langle \phi(x_i), v \rangle < 0$ on the set $\mathcal{X}_-$. These equations say that the points in the two sets $\phi(\mathcal{X}_-)$ and $\phi(\mathcal{X}_+)$ are separated by a hyperplane determined by v.

Thus, by using a feature map, our problem in shape recognition is reduced to a question about determining when two sets of points in a Hilbert space are separated by a hyperplane, and finding such a hyperplane.

If there is a hyperplane that separates $\phi(\mathcal{X}_-)$ from $\phi(\mathcal{X}_+)$, then it is not hard to see that there are infinitely many such hyperplanes.

So, a natural question to ask is: can we pick a "best" hyperplane? One special hyperplane gives rise to what is known as the $\boxed{\text{MAXIMAL MARGIN CLASSIFIER}}$.

In the next section, we study this hyperplane separation problem and introduce this special hyperplane.

8.4 The maximal margin classifier

If $\mathcal{H}$ is a real Hilbert space, then a $\boxed{\text{HYPERPLANE}}$ in $\mathcal{H}$ is a closed affine subspace of codimension 1. Every hyperplane is of the form $V = \{x \in H : \langle x, v \rangle = c\}$ for some constant c and some nonzero vector v. There is some redundancy in the definition. In particular, if $c = 0$ we may replace v by rv for any nonzero scalar. When $c \neq 0$ then we may replace the pair (v, c) by (rv, rc) for any nonzero scalar r. But up to this change the hyperplanes are in one-to-one correspondence with all such pairs of vectors and constants. If we require v to be a unit vector, then the only pairs are (v, c) and $(-v, -c)$.

Any hyperplane $V = \{x \in \mathcal{H} : \langle x, v \rangle = c\}$ partitions the space $\mathcal{H}$ into three sets: the hyperplane V and two open subsets called the $\boxed{\text{SIDES OF THE HYPERPLANE}}$ given by the equations $V_+ = \{x \in \mathcal{H} : \langle x, v \rangle > c\}$ and $V_- = \{x \in \mathcal{H} : \langle x, v \rangle < c\}$.

Definition 8.1. Let $\mathcal{X} = \{x_i : i \in I\} \subset \mathcal{H}$ be a data set and let $\lambda_i \in \{\pm 1\}$ be labels. Let $\mathcal{X}_+ = \{x_i : \lambda_i = +1\}$ and $\mathcal{X}_- = \{x_i : \lambda_i = -1\}$. The data is said to be $\boxed{\text{LINEARLY SEPARABLE}}$ if there is a hyperplane V such that $\mathcal{X}_+ \subseteq V_+$ and $\mathcal{X}_- \subseteq V_-$. Otherwise the data is said to be $\boxed{\text{NON-SEPARABLE}}$.

For the moment we will put the RKHS viewpoint aside and consider just the hyperplane separation problem for an arbitrary Hilbert space. We will then return to the function space viewpoint at the end.

Given a linearly separable data set $\mathcal{X}$ the problem is to find an affine hyperplane that separates the two classes. The geometric approach we will describe now is called the $\boxed{\text{MAXIMAL MARGIN HYPERPLANE}}$. This approach has the appealing property that the normal vector to the hyperplane lies in the span of the points x_i. Once again the numerical problem that appears at the end will depend only on the inner products $\langle x_i, x_j \rangle$.

We need some preliminary results. Recall that, given a subset $V \subseteq \mathcal{H}$, by the distance from x to V we mean

$$d(x, V) = \inf\{\|x - y\| : y \in V\}.$$

Lemma 8.2. *Let* $V = \{y \in \mathcal{H} : \langle y, v \rangle = c\}$ *be an affine hyperplane. Then the distance of a point x from V is given by*

$$d(x, V) = \frac{|\langle x, v \rangle - c|}{\|v\|}.$$

Proof. We first translate the hyperplane, so that it passes through the origin. This translated hyperplane is $V' = \{y \in \mathcal{H} : \langle y, v \rangle = 0\}$. Set $k = c/\|v\|^2$. If

$y \in V$, then $\langle y - kv, v \rangle = \langle y, v \rangle - c = 0$ so that $y - kv \in V'$. Conversely, if $y \in V'$ then $y + kv \in V$. Thus, we see that $V' = V - kv$.

The hyperplane V' is a subspace of H. The distance of a point h from this subspace is the length of the projection of h onto the orthogonal complement of V'. Since the orthogonal complement is spanned by the vector v the projection is given by $P(h) = \frac{\langle h, v \rangle}{\|v\|^2} v$. Hence the length of the projection is $\|P(h)\| = |\langle h, v \rangle| / \|v\|$.

Since translation preserves distance, for any $x \in \mathcal{H}$ we have that

$$d(x, V) = d(x - kv, V - kv) = d(x - kv, V') = \|P(x - kv)\| = \frac{|\langle x, v \rangle - c|}{\|v\|}.$$

$\square$

The next result shows that when two finite sets of vectors can be separated by a hyperplane, then the vector that determines the separating hyperplane can always be chosen in the span of the vectors. In addition, this can be done without altering the distance to the hyperplane.

Proposition 8.3. *Let* $\mathcal{X} = \{x_{(i)} : i = 1, \ldots, n\}$ *be a set of points in a real Hilbert space* $\mathcal{H}$. *Let* $\mathcal{X}_+$, $\mathcal{X}_-$ *be a partition of* $\mathcal{X}$ *into two disjoint subsets. If* $\mathcal{X}_+$ *and* $\mathcal{X}_-$ *are linearly separable by a hyperplane of the form* $V = \{x : \langle x, v \rangle = c\}$, *then the vector* v *can be chosen to be an element of the span of* $\mathcal{X}$.

Proof. Let P denote the orthogonal projection onto the span of $\mathcal{X}$. We have $\langle x_i, v \rangle = \langle Px_i, v \rangle = \langle x_i, Pv \rangle$. Therefore, the vector Pv separates the classes $\mathcal{X}_-$ and $\mathcal{X}_+$. $\square$

It is always possible to embed a set of points into a sufficiently high-dimensional space in such a way that the points are linearly separable. The following result gives us information about the dimension of the space that is needed.

Proposition 8.4. *Let* $\mathcal{X} = \{x_i : i = 1, \ldots, n\}$ *be a set of distinct points in a real Hilbert space* $\mathcal{H}$. *Let* W *denote the span of the vectors in* $\mathcal{X}$.

If $\dim(W) = n$, *and* $\mathcal{X}_+$ *and* $\mathcal{X}_-$ *are any partition of* $\mathcal{X}$ *into two disjoint subsets, then there exists a hyperplane that separates* $\mathcal{X}_+$ *and* $\mathcal{X}_-$.

On the other hand, if $\dim(W) \leq n - 2$, *then there exists a partition of* $\mathcal{X}$ *into two disjoint subsets,* $\mathcal{X}_+$ *and* $\mathcal{X}_-$, *such that no affine hyperplane separates* $\mathcal{X}_+$ *and* $\mathcal{X}_-$.

Proof. After re-indexing the points we can assume that $\mathcal{X}_+ = \{x_i \ : \ i \le m\}$. We will first prove the case where $\dim(W) = n$. In this case we will show that there is a hyperplane through the origin that separates the two classes.

Let $\lambda_i = \pm 1$ depending on whether $x_i \in \mathcal{X}_\pm$. We begin by choosing a dual basis for the points x_i as follows. Since the points x_i are linearly independent we can choose a dual basis for the space W: that is, a set of n vectors $v_1, \ldots, v_n$ such that $\langle x_i, v_j \rangle = \delta_{i,j}$. Now let $v = \sum_{j=1}^n \lambda_j v_j \in W$. We have

$$\langle x_i, v \rangle = \sum_{j=1}^n \lambda_j \langle x_i, v_j \rangle = \lambda_i.$$

Therefore the hyperplane $V = \{x \in H \ : \ \langle x, v \rangle = 0\}$ separates $\mathcal{X}_+$ and $\mathcal{X}_-$.

To prove the theorem in the case where $\dim(W) \le n - 2$ we first apply the map $\phi(x) = x \oplus 1$ which embeds $\mathcal{H}$ into the Hilbert space $\mathcal{H} \oplus \mathbb{R}$. The dimension of the vector space spanned by the vectors $\{\phi(x_i)\}$ is at most $n - 1$. Hence, there exist $\alpha_1, \ldots, \alpha_n$, not all 0, such that $\sum_{i=1}^n \alpha_i (x_i \oplus 1) = 0$. It follows that $\sum_{i=1}^n \alpha_i = 0$ and that $\sum_{i=1}^n \alpha_i x_i = 0$.

We re-index the points x_i so that the $\alpha_1, \ldots, \alpha_m > 0$ and $\alpha_{m+1}, \ldots, \alpha_n \le 0$. Note that there is at least one i for which $\alpha_i < 0$ and one index j for which $\alpha_j > 0$. Let $\mathcal{X}_+ = \{x_i \ : \ i \le m\}$ and $\mathcal{X}_- = \{x_i \ : \ i \ge m + 1\}$.

Now if a hyperplane given by (v, c) separated this set, then we would have $\sum_{i=1}^m \alpha_i \langle x_i, v \rangle = -\sum_{i=m+1}^n \alpha_i \langle x_i, v \rangle$ and $\sum_{i=1}^m \alpha_i = \sum_{i=m+1}^n -\alpha_i$. Multiplying by c and subtracting we get

$$\sum_{i=1}^m \alpha_i (\langle x_i, v \rangle - c) = \sum_{i=m+1}^n -\alpha_i (\langle x_i, v \rangle - c).$$

The left-hand side is a sum of nonnegative terms, while the right-hand side is a sum of nonpositive terms. So the only way that equality can occur is if all the terms are 0. Since the hyperplane is separating, $\langle x_i, v \rangle \ne c$ for all i and so $\alpha_i = 0$ for all i. This contradicts the fact that there is an index for which $\alpha_i \ne 0$.

Hence, there can be no separating hyperplane. $\square$

The reader will notice that the above theorem says nothing about the case where $\dim(W) = n - 1$. A set of n points that span an $(n - 1)$-dimensional space may or may not be separable by an affine hyperplane.

We have established two facts. First, if the data set is linearly independent, then it can be linearly separated by a hyperplane through the origin. Second, the vector that determines the hyperplane can be chosen in the span of the vectors x_i.

Thus, given a finite set $\{x_1, \ldots, x_n\} \subset X$ by choosing a kernel function K on X such that the kernel functions $k_{x_1}, \ldots, k_{x_n}$ are linearly independent we can always embed the data set in a way that makes it linearly separable. In particular, if we choose a feature map $\phi : X \to \mathcal{H}$ such that the vectors $\phi(x_i)$, $i = 1, \ldots, n$ are linearly independent, then the pull-back kernel will have this property.

Definition 8.5. Suppose that $\mathcal{X} \subseteq \mathcal{H}$ has two classes $\mathcal{X}_+$ and $\mathcal{X}_-$ that can be separated by a hyperplane. The $\boxed{\text{MAXIMAL MARGIN HYPERPLANE}}$ is the hyperplane that minimizes the maximal distance of the points from the hyperplane V.

There are many ways to see that such a hyperplane exists. We leave a geometrical proof to the exercises and give here an optimization proof.

Theorem 8.6. *Suppose that $\mathcal{X}$ is a subset of a real Hilbert space $\mathcal{H}$ with two classes $\mathcal{X}_+$ and $\mathcal{X}_-$ which can be linearly separated. Consider the problem of minimizing $\frac{1}{2}\|v\|^2$ subject to the constraints $\lambda_i(\langle x_i, v\rangle - c) \geq 1$. Then there is a vector w that attains this minimum, and moreover w is unique, w is in the span of the vectors $x_1, \ldots, x_n$ and the hyperplane determined by (w, c) is a maximal margin hyperplane.*

Proof. Note that if (v, c) determines a separating hyperplane V, then (rv, rc) determines the same hyperplane. Thus, replacing (v, c) by (rv, rc) we may always assume that $|\langle x_i, v\rangle - c| \geq 1$ for every i, with equality for at least one i_0. In this case, $d(x_i, V) \geq \frac{1}{\|v\|}$ and $d(x_{i_0}, V) = \frac{1}{\|v\|}$.

Thus, the minimal distance is maximized when $\frac{1}{\|v\|}$ is maximized, or when $\|v\|$ is minimized subject to the constraints $\lambda_i(\langle x_i y, v\rangle - c) \geq 1$ for $i = 1, \ldots, n$.

We first discuss uniqueness. The set of points v for which there exists c so that (v, c) satisfies the constraints $\lambda_i(\langle x_i, v\rangle - c) \geq 1$ is convex and closed. It is a well-known result in Hilbert space that closed convex sets contain a unique point of smallest norm.

Thus, the maximal margin hyperplane exists and is determined by (w, c), where $\|w\|$ is minimized.

Let P denote the projection onto the span of the vectors $x_1, \ldots, x_n$. We have $\|Pw\| \leq \|w\|$. In addition, we have $\langle x_i, Pw\rangle = \langle Px_i, w\rangle = \langle x_i, w\rangle$. It follows that the point Pw also solves the minimization problem above. Hence, from the uniqueness assertion we see that $Pw = w$ and so w is in the span of the vectors $v_1, \ldots, v_n$.

Finally, observe that there is no difference between minimizing $\|v\|$ or $\frac{1}{2}\|v\|^2$. □

This theorem shows that the unique solution to the above minimization problem is of the form $w = \sum_{j=1}^{n} \alpha_j x_j$. We can therefore assume at the outset that the minimization is not over all $v \in \mathcal{H}$ but just over v in the span of $x_1, \ldots, x_n$. This is a finite-dimensional space, and so our problem is finite-dimensional.

Suppose that $v = \sum_{j=1}^{n} \alpha_j x_j$. Then $\|v\|^2 = \sum_{i,j=1}^{n} \alpha_i \alpha_j \langle x_j, x_i \rangle = \langle Q\alpha, \alpha \rangle$ where Q is the positive matrix $\langle x_j, x_i \rangle$. The constraints are now of the form $\lambda_i (\sum_{j=1}^{n} \alpha_j \langle x_j, x_i \rangle - c) \geq 1$. The problem is to find α and c that minimize $\langle Q\alpha, \alpha \rangle$ subject to these constraints.

These observations show us that we only need to store the inner products $\langle x_j, x_i \rangle$. Computationally, this optimization problem is quite easily solved using techniques from quadratic programming.

8.5 Summary: a functional perspective

We now give an RKHS interpretation of the maximal margin classifier and shape recognition. This description is very similar to the problems we looked at in the chapter on interpolation and approximation.

First we are given a data set $\{x_i; 1 \leq i \leq n\} \subseteq \mathcal{X}$ and labels λ_i which partition the data into $\mathcal{X}_+$, the "outs", and $\mathcal{X}_-$, the "ins", and we wish to build a function that will predict if some other point in $\mathcal{X}$ is "in" or "out". We apply a feature map $\phi : X \to \mathcal{L}$ to create an RKHS $\mathcal{H}$ on X and identify $x \mapsto k_x(\cdot) = \langle \cdot, \phi(x) \rangle$. Without loss of generality we can add an affine term and embed $k_x \in \mathcal{H}$ as $k_x \oplus 1 \in \mathcal{H} \oplus \mathbb{R}$. The advantage of this new embedding is that if the points k_{x_i} are linearly separable in the space $\mathcal{H}$, say by a vector v and constant c, then the points $k_{x_i} \oplus 1$ are linearly separable in the space $\mathcal{H} \oplus \mathbb{R}$ by the hyperplane through the origin determined by the vector $(v, -c)$.

Note that our original map ϕ determined the kernel $K(x, y) = \langle \phi(x), \phi(y) \rangle$ and our new map $\psi : X \to \mathcal{H} \oplus \mathbb{C}$ given by $\psi(x) = k_x \oplus 1$ determines the new kernel

$$K_1(x, y) = \langle \psi(x), \psi(y) \rangle = \langle k_x, k_y \rangle + 1 = K(y, x) + 1 = K(x, y) + 1,$$

since we are in the real case.

Thus, we may assume that our feature map ϕ gives rise to a situation where the kernel functions k_{x_i} are linearly separable by a hyperplane through the origin.

With this change, to find the maximal margin classifier, the problem is to minimize $\|f\|^2$ over all $f \in \mathcal{H}(K)$ subject to the constraints that $\lambda_i \langle f, k_{x_i} \rangle =$

$\lambda_i f(x_i) \geq 1$. This is similar to the interpolation problem studied in Chapter 3. We need to find a function $f \in \mathcal{H}(K)$ of smallest norm that satisfies certain inequalities, instead of equalities, at prescribed points.

The function f acts as our "predictor". When $f(x) < 0$ then we predict that x is in the set, and if $f(x) > 0$ then we predict that x is outside the set.

Each time that an oracle gives us a new point x_{n+1} and label λ_{n+1}, we can re-compute a new maximal margin classifier $\hat{f}$ for this improved data set. In this way the predictor "learns" and improves its predictions.

Now once we have chosen and fixed the feature map ϕ there is no guarantee that the enlarged data set and labels is still linearly separable by a hyperplane.

Given any real RKHS $\mathcal{H}$ on X, points $x_1, \ldots, x_n$ in X and labels $\lambda_i \in \{\pm 1\}$, there will exist a function $f \in \mathcal{H}$ satisfying $\lambda_i f(x_i) \geq 1$ if and only if $k_{x_1}, \ldots, k_{x_n}$ is linearly separable by a hyperplane through the origin.

So we would next like to study how to still choose a suitable function f in the non-separable case.

We can recast this problem in a way that resembles the regression problem. Note that a function f satisfies $\lambda_i f(x_i) \geq 1$ if and only if $1 - \lambda_i f(x_i) \leq 0$. In the case that there is no separating hyperplane, there will be no such functions. A heuristic solution is to minimize a loss function L that penalizes functions f that fail to satisfy the above inequality. Let $L(x, y, f(x)) = \max\{0, 1 - yf(x)\}$. Note that any function f that satisfies the above inequalities has the property that $L(x_i, \lambda_i, f(x_i)) = 0$.

In the non-separable case we attempt to minimize $J(f) = \sum_{i=1}^n \max\{0, 1 - \lambda_i f(x_i)\}$ over all functions $f \in \mathcal{H}$. Note that in the linearly separable case the minimum is 0, otherwise we construct a function that, on the average, least violates the constraints.

Once again the reason this new problem is tractable is that the loss function depends only on the labels and the values of the function at the points x_i. A theorem, known as the representer theorem, guarantees that the solution lies in the span of $k_{x_1}, \ldots, k_{x_n}$. This is the result we look at next.

8.6 The representer theorem

Many results in statistics involve functional approximation and the use of a loss or penalty function. In this sense the solution to problems in statistics are extremal solutions to problems of the type

$$\min_{f \in H} W(\|f\|_{\mathcal{H}}^2) + L(f(x_1), \ldots, f(x_n)),$$

where $\mathcal{H}$ is a Hilbert space. For example we saw that the maximal margin classifier can be replaced by the problem of minimizing the loss function $L(f(x_1), \ldots, f(x_n)) = \sum_{i=1}^{n} \max\{0, 1 - \lambda_i f(x_i)\}$.

The problem with this loss function is that the function f that minimizes this quantity tends to overfit the data. It performs well on the existing data, but tends to misclassify new data. To offset this problem a term of the form $\| f \|_{\mathcal{H}}^2$ is added to the loss function, to penalize functions of large norm.

The more general problem is to minimize $W(\| f \|_H^2) + L(f(x_1), \ldots, f(x_n))$ where W is increasing and L is a loss function.

A priori there is no reason why such a problem should be tractable. However, under reasonable assumptions on the functions W and L, all solutions to the above problem are of the form $f = \sum_{i=1}^{n} \alpha_i k_{x_i}$, i.e. in the span of the kernel functions.

This tells us two things. First, the problem is intrinsically finite-dimensional. We do not need to leave the span of the functions $k_{x_1}, \ldots, k_{x_n}$ in order to solve the problem. Second, we know that all the information about this span is contained in the values of the kernel function for the observations $x_1, \ldots, x_n$, i.e. by the numbers $K(x_i, x_j)$.

This result goes by the rather uninformative name of the "representer theorem".

Theorem 8.7. *Let $\mathcal{H}$ be an RKHS on X. Let $W : \mathbb{R} \to \mathbb{R}$ be a monotonically increasing function and let $L : \mathbb{R}^n \to \mathbb{R}$ be continuous. Consider the cost function $J(f) = W(\| f \|_{\mathcal{H}}^2) + L(f(x_1), \ldots, f(x_n))$.*

If f^* is a function such that $J(f^*) = \inf_{f \in \mathcal{H}} J(f)$, then f^* is in the span of the functions $k_{x_1}, \ldots, k_{x_n}$.

Proof. Let S denote the span of the kernel functions at the points $x_1, \ldots, x_n$ and write $f^* = g + h$, where $g \in S$ and $h \perp S$. The claim is that $h = 0$.

Since $h \in S^{\perp}$ $h(x_i) = 0$ for all i and since the function L only depends on the values of the function at the points x_i we see that $J(f^*) = W(\| h \|^2 + \| g \|^2) + L(g(x_1), \ldots, g(x_n))$. Furthermore, since W is monotonically increasing we see that $J(f^*) > W(\| g \|^2) + L(g(x_1), \ldots, g(x_n)) = J(g)$. It follows that $f^* \in S$. $\qquad\square$

This theorem is simple, yet remarkable. The initial problem is stated as an optimization problem in the RKHS $\mathcal{H}$. The space $\mathcal{H}$ could be, and often is, infinite-dimensional. The theorem shows that any solution to this problem must be in the finite-dimensional space spanned by $k_{x_1}, \ldots, k_{x_n}$.

For the special case where the first term W is of the form $C \, \|f\|^2$ for some constant C and the second term L is convex we can actually show that the solution exists and is unique.

Recall that a real-valued function L defined on a convex set C is called a CONVEX FUNCTION if for $x, y \in C$ and $0 < t < 1$, we have that $L(tx + (1-t)y) \le tL(x) + (1-t)L(y)$.

Theorem 8.8. *Consider the problem of minimizing the function* $J(f) = \|f\|_H^2 + L(f(x_1), \ldots, f(x_n))$ *where L is convex. The solution to this problem exists and is unique.*

Proof. The uniqueness of the solution uses an argument that appears frequently in Hilbert space. For simplicity of notation we shall denote $L(f(x_1), \ldots, f(x_n))$ by $L(f)$. Suppose that f, g are minimizers of the above problem, then $\|f\|^2 + L(f) \le \left\|\frac{f+g}{2}\right\|^2 + L(\frac{1}{2}(f+g))$. Since g is also a minimizer we have the same inequailty when we replace f by g. Adding these two inequalities and dividing by 2 we get

$$\frac{1}{2}(\|f\|^2 + \|g\|^2) - \left\|\frac{1}{2}(f+g)\right\|^2 \le L(\frac{1}{2}(f+g)) - \frac{1}{2}L(f) - \frac{1}{2}L(g).$$

From the parallelogram law we can write $\left\|\frac{1}{2}(f-g)\right\|^2 = -\left\|\frac{1}{2}(f+g)^2\right\| + 2\left\|\frac{1}{2}f\right\|^2 + 2\left\|\frac{1}{2}g\right\|^2$. Hence, the left-hand side of this expression is $\left\|\frac{1}{2}(f-g)\right\|^2$. On the other hand we assumed that L is convex and so $L(\frac{f+g}{2}) \le \frac{1}{2}L(f) + \frac{1}{2}L(g)$. Combining these facts we get $\|f - g\|^2 \le 0$ and so $f - g$. $\square$

Note that the function $\max\{0, 1 - \lambda_i f(x_i)\}$ is a convex function of f. Indeed, we have $\max\{0, t\alpha + (1-t)\beta\} \le t\max\{0, \alpha\} + (1-t)\max\{0, \beta\}$. From which this follows, setting $\alpha = 1 - \lambda_i f(x_i)$ and $\beta = 1 - \lambda_i g(x_i)$ we get $t\alpha + (1-t)\beta = 1 - \lambda_i(tf + (1-t)g)(x_i)$.

Thus, the two representer theorems tell us that if we choose a loss/penalty function of the form $J(f) = W(\|f\|) + \sum_{i=1}^{n} \max\{0, 1 - \lambda_i f(x_i)\}$ with W monotonically increasing, then there will always exist a unique function in $\mathcal{H}(K)$ that minimizes J and that function will be in the span of the kernels $k_{x_1}, \ldots, k_{x_n}$.

Thus, once we have fixed the RKHS of functions on the set X, for any finite data set and labels we will be able to solve a finite dimensional problem to find a function f such that $f(x) < 0$ "predicts" belonging to our set S and which minimizes our loss function.

8.7 Exercises

Exercise 8.1. Let $\mathcal{X}_+$ and $\mathcal{X}_-$ be two finite sets in a real Hilbert space that can be separated by a hyperplane. Let C_+ and C_- be their closed convex hulls. By a standard result in Hilbert space, there is a pair of points, $y_+ \in C_+$ and $y_- \in C_-$ such that $\|y_+ - y_-\| = \inf\{\|w_+ - w_-\| : w_+ \in C_+, w_- \in C_-\}$. Prove that $V = \{x : \langle x, y_+ - y_- \rangle = \|y_+\|^2 - \|y_-\|^2\}$ is a maximal marginal hyperplane.

Exercise 8.2. Suppose that $\mathcal{X}$ is a linearly separable data set with two classes, $\mathcal{X}_+$ and $\mathcal{X}_-$. If $V = \{x \in \mathcal{H} : \langle x, v \rangle = c\}$ is a maximal margin hyperplane, then prove that $d(\mathcal{X}_+, V) = d(\mathcal{X}_-, V)$.

Exercise 8.3. Let K be a kernel function of the form $K(x, y) = \langle \phi(x), \phi(y) \rangle$ where $\phi : X \to \mathcal{H}$ and $\mathcal{H}$ is finite-dimensional. Suppose that the dimension of $\mathcal{H}$ is m. Prove that for any set of points $\{x_1, \ldots, x_n\}$ the rank of the matrix $\big(K(x_i, x_j)\big)$ is at most m.

9

Negative definite functions

In this chapter we study the theory of conditionally negative definite functions. These functions arise in many places in analysis and are often a convenient way to obtain kernel functions. In particular, Schoenberg developed this theory and then used it to characterize the metric spaces that can be embedded isometrically into a Hilbert space. Negative definite functions also play a major role in group representation theory, which will be the topic of the next chapter.

9.1 Schoenberg's theory

Definition 9.1. A function $\psi : X \times X \to \mathbb{C}$ is called $\boxed{\text{CONDITIONALLY}}$ $\boxed{\text{NEGATIVE DEFINITE}}$ if and only if, given $x_1, x_2, \ldots, x_n \in X$ and $\alpha_1, \ldots, \alpha_n \in \mathbb{C}$ such that $\sum_{j=1}^{n} \alpha_j = 0$, we have

$$\sum_{i,j=1}^{n} \overline{\alpha_i} \alpha_j \psi(x_i, x_j) \leq 0.$$

The function ψ is called $\boxed{\text{NEGATIVE DEFINITE}}$ provided that this latter inequality is met for all $\alpha_1, \ldots, \alpha_n \in \mathbb{C}$.

Clearly, ψ is negative definite if and only if $-\psi$ is positive definite, i.e. a kernel function. Again, some caution is needed: many other authors use the term "negative definite function" to refer to functions that we are calling conditionally negative definite.

Note that if c is any complex number, and we consider the constant function $\psi(x, y) = c$, then ψ is conditionally negative definite. Thus, in general, for a conditionally negative definite function, $\psi(x, x)$ need not be real. Also, if we choose c so that $c \neq \overline{c}$, then $\psi(y, x) \neq \overline{\psi(x, y)}$.

Definition 9.2. A conditionally negative definite function $\psi : X \times X \to \mathbb{C}$ is called $\boxed{\text{SELF-ADJOINT}}$ provided that $\psi(y, x) = \overline{\psi(x, y)}$ and $\boxed{\text{SYMMETRIC}}$ provided that $\psi(y, x) = \psi(x, y)$ for all $x, y \in X$.

When ψ is real-valued these two definitions coincide.

It is easily seen that the sum of (conditionally) negative definite functions is (conditionally) negative definite, and the pointwise limit of a sequence of (conditionally) negative definite functions is also (conditionally) negative definite.

We now show how to construct a kernel function from a conditionally negative definite function ψ.

Proposition 9.3. *Let* $\psi : X \times X \to \mathbb{C}$, *fix a point* $x_0 \in X$ *and set* $K(x, y) = -\psi(x, y) + \psi(x, x_0) + \psi(x_0, y) - \psi(x_0, x_0)$. *The function* ψ *is conditionally negative definite if and only if* K *is a kernel function.*

Proof. Assume that K is a kernel function and assume $\sum_{j=1}^{n} \alpha_j = 0$. We have

$$0 \le \sum_{i,j=1}^{n} \overline{\alpha_i} \alpha_j K(x_i, x_j)$$

$$= - \sum_{i,j=1}^{n} \bar{\alpha}_i \alpha_j \psi(x_i, x_j) + \sum_{i,j=1}^{n} \bar{\alpha}_i \alpha_j \psi(x_i, x_0)$$

$$+ \sum_{i,j=1}^{n} \bar{\alpha}_i \alpha_j \psi(x_0, x_j) - \sum_{i,j=1}^{n} \bar{\alpha}_i \alpha_j \psi(x_0, x_0)$$

$$= - \sum_{i,j=1}^{n} \bar{\alpha}_i \alpha_j \psi(x_i, x_j).$$

This establishes the fact that ψ is conditionally negative definite. Now suppose that ψ is conditionally negative definite and that $\alpha_1, \alpha_2, \dots, \alpha_n \in \mathbb{C}$. Let $\alpha_0 = - \sum_{j=1}^{n} \alpha_j$. It follows that $\sum_{j=0}^{n} \alpha_j = 0$. We have

$$\sum_{i,j=1}^{n} \overline{\alpha_i} \alpha_j K(x_i, x_j)$$

$$= - \sum_{i,j=1}^{n} \overline{\alpha_i} \alpha_j \psi(x_i, x_j) + \sum_{i,j=1}^{n} \overline{\alpha_i} \alpha_j \psi(x_i, x_0)$$

$$+ \sum_{i,j=1}^{n} \overline{\alpha_i} \alpha_j \psi(x_0, x_j) - \sum_{i,j=1}^{n} \overline{\alpha_i} \alpha_j \psi(x_0, x_0)$$

$$= -\sum_{i,j=1}^{n} \overline{\alpha_i}\alpha_j \psi(x_i, x_j) + \Big(\sum_{j=1}^{n}\alpha_j\Big)\Big(\sum_{i=1}^{n}\overline{\alpha_i}\psi(x_i, x_0)\Big)$$

$$+ \Big(\sum_{i=1}^{n}\overline{\alpha_i}\Big)\Big(\sum_{j=1}^{n}\alpha_j\psi(x_0, x_j)\Big) - |\sum_{i=1}^{n}\alpha_i|^2\psi(x_0, x_0)$$

$$= -\sum_{i,j=1}^{n}\overline{\alpha_i}\alpha_j\psi(x_i, x_j) + \big(-\alpha_0\big)\Big(\sum_{i=1}^{n}\overline{\alpha_i}\psi(x_i, x_0)\Big)$$

$$+ \big(-\overline{\alpha_0}\big)\Big(\sum_{j=1}^{n}\alpha_j\psi(x_0, x_j)\Big) - (\overline{\alpha_0}\alpha_0)\psi(x_0, x_0)$$

$$= -\sum_{i,j=0}^{n}\overline{\alpha_i}\alpha_j\psi(x_i, x_j) \geq 0,$$

since ψ is conditionally negative definite. This shows that K is a kernel function. $\qquad\square$

We call a conditionally negative definite function $\boxed{\text{NORMALIZED}}$ if $\psi(x, x) = 0$ for all $x \in X$.

Proposition 9.4. *If $\mathcal{H}$ is a Hilbert space, then the function $\psi(x, y) = \|x - y\|^2$ is a self-adjoint, normalized, conditionally negative definite function.*

We leave the proof as an exercise. If we choose the base point in the Hilbert space to be $x_0 = 0$, then the induced kernel function is $K(x, y) = -\|x - y\|^2 + \|x\|^2 + \|y\|^2 = \langle x, y \rangle + \langle y, x \rangle = 2Re(\langle x, y \rangle)$. The function $\|x - y\|^2$ is the prototypical conditionally negative definite function in much the same way that the Hilbert space inner product is the prototypical kernel function. This statement is made precise by the following result of Schoenberg, often known as $\boxed{\text{SCHOENBERG'S LEMMA}}$, since he only considered it as a preliminary to the result which follows. Note that Schoenberg's lemma also provides a converse to the last proposition.

Theorem 9.5 (Schoenberg's lemma). *Let $\psi : X \times X \to \mathbb{C}$ be a symmetric, normalized conditionally negative definite function. Then there exists a Hilbert space $\mathcal{H}$ and a map $\phi : X \to \mathcal{H}$ such that $\psi(x, y) = \|\phi(x) - \phi(y)\|^2$. In particular, ψ must be real-valued.*

Proof. Fix any $x_0 \in X$ and let K be the kernel function from Proposition 9.3. Consider the map $\phi : X \to \mathcal{H}(K)$ given by $\phi(x) = \frac{1}{\sqrt{2}} k_x$. We have

$$\|\phi(x) - \phi(y)\|^2 = \frac{1}{2} \|k_x - k_y\|^2$$

$$= \frac{1}{2} (K(x, x) - K(x, y) - K(y, x) + K(y, y))$$

$$= \frac{1}{2} (\psi(x, y) + \psi(y, x)) = \psi(x, y),$$

after expanding and using the fact that $\psi(x, x) = \psi(y, y) = \psi(x_0, x_0) = 0$. $\square$

A corollary of the above result is the following result of Schoenberg, which characterizes when a metric space can be embedded into a Hilbert space.

Theorem 9.6 ($\boxed{\text{SCHOENBERG'S EMBEDDING THEOREM}}$). *Let (X, d) be a metric space. Then there exists a Hilbert space $\mathcal{H}$ and an isometry $\phi : X \to \mathcal{H}$ if and only if d^2 is conditionally negative definite.*

Proof. Since d is a metric, we see that $d^2(x, y) = d^2(y, x)$ and $d^2(x, x) = 0$ for all x. The result now follows from the previous theorem. $\square$

Given a self-adjoint conditionally negative definite function ψ it is possible to construct two families of kernel functions. The first is obtained by exponentiating, to get $K_t(x, y) = e^{-t\psi(x,y)}$. The second is given by $L_t(x, y) = \dfrac{t}{t + \psi(x, y)}$. The fact that these are all kernel functions is the subject of the next two results.

Theorem 9.7 ($\boxed{\text{SCHOENBERG'S GENERATOR THEOREM}}$). *The function $\psi : X \times X \to \mathbb{C}$ is a self-adjoint conditionally negative definite function if and only if the function $K_t = e^{-t\psi}$ is a kernel function for all $t > 0$.*

Proof. First let us assume that the function $K_t(x, y) = e^{-t\psi(x,y)}$ is a kernel function for all $t > 0$. It follows that $-K_t$ is negative definite and so $1 - K_t$ is self-adjoint and conditionally negative definite. Thus, the function $\gamma_t = \dfrac{1 - K_t}{t}$ is self-adjoint and conditionally negative definite.

Since limits of self-adjoint conditionally negative definite functions are self-adjoint and conditionally negative definite, we have that

$$\lim_{t \to 0} \frac{1 - e^{-t\psi(x,y)}}{t} = \psi(x, y)$$

is self-adjoint and conditionally negative definite.

Now assume that ψ is self-adjoint and conditionally negative definite. Since $t\psi$ is also self-adjoint and conditionally negative definite for any $t > 0$, to prove that $e^{-t\psi}$ is a kernel function for all $t > 0$, it will be enough to show that $K_1 = e^{-\psi}$ is a kernel function.

Now fix any point x_0 and let J be the kernel function given in Proposition 9.3. By Theorem 4.16 we have that e^J is a kernel function.

We have

$$e^{-\psi(x,y)} = e^{\left(\psi(x_0,x_0)-\psi(x_0,y)-\psi(x,x_0)+J(x,y)\right)}$$
$$= e^{\psi(x_0,x_0)}e^{-\psi(x,x_0)}e^{J(x,y)}e^{-\psi(x_0,y)}.$$

Since $\psi(x_0, x_0) \in \mathbb{R}$, in the above expression the first factor is a positive constant. The third factor is a kernel. By Exercise 2.2 and the fact that ψ is self-adjoint, the product of the last three terms is a kernel function. Hence, the product of all four terms is a kernel function, i.e. $e^{-\psi(x,y)}$ is a kernel function. $\square$

Note that the functions K_t satisfy $K_t \cdot K_s = K_{t+s}$. Thus, we have a semigroup of kernel functions and Schoenberg's generator theorem characterizes the functions that generate such a semigroup of kernels via exponentiation. We will look at this idea more deeply in Section 9.2 on infinitely divisible kernels.

Corollary 9.8. *Let $\mathcal{H}$ be a Hilbert space. Then $K_t(x, y) = e^{-t\|x-y\|^2}$ is a kernel function for all $t > 0$.*

The above kernel function is known as the $\boxed{\text{GAUSSIAN KERNEL}}$.

If ψ is a conditionally negative definite function such that $\psi(x, y) \geq 0$ for all $x, y \in X$, then there is another family of kernels associated to ψ.

Theorem 9.9. *Let $\psi : X \times X \to \mathbb{C}$ with $Re(\psi(x, y)) \geq 0$ for all $x, y \in X$. Then ψ is self-adjoint and conditionally negative definite if and only if the function $L_t = \dfrac{t}{t + \psi}$ is a kernel function for all $t > 0$.*

Proof. The proof is interesting as it illustrates the importance of integral transforms. We saw earlier (Exercise 2.6) that the Fourier transform can be used to produce positive definite functions on the line. We use the Laplace transform to prove this result.

If $f : \mathbb{R}_+ \to \mathbb{C}$ is a continuous function, then the Laplace transform of f is the function F defined by $F(z) = \int_0^\infty f(t)e^{-zt} dt$. Under suitable assumptions on f the Laplace transform converges and F is well-defined. It is an easy calculation that the Laplace transform of the function $f(t) = e^{-t}$ is the function $F(x) = \frac{1}{1+x}$.

With these preliminaries out of the way we prove our theorem. Note that $\lim_{t\to\infty} t(1 - L_t) = \psi$, pointwise.

Thus, if L_t is a kernel for all $t > 0$, then the function $1 - L_t$ is self-adjoint and conditionally negative definite and so ψ is self-adjoint and conditionally negative definite. This establishes one implication.

Conversely, if ψ is self-adjoint and conditionally negative definite, then so is ψ/t for $t > 0$. Since $L_t = \frac{1}{1+\psi/t}$, to prove that L_t is a kernel for all $t > 0$, it will be enough to prove that $L_1 = \frac{1}{1+\psi}$ is a kernel function. Let $\alpha_1, \ldots, \alpha_n \in \mathbb{C}$ and let $x_1, \ldots, x_n \in X$, we have

$$\sum_{i,j=1}^{n} \overline{\alpha_i}\alpha_j L_1(x_i, x_j) = \int_0^\infty e^{-t} \left(\sum_{i,j=1}^{n} \overline{\alpha_i}\alpha_j e^{-t\psi(x_i,x_j)} \right) dt$$

Since by assumption $Re(\psi(x, y)) \geq 0$, it is easily seen that the above integral does converge. The previous theorem shows that the integrand is a nonnegative function of t, since $e^{-t\psi}$ is a kernel. It follows that the integral is nonnegative and hence L_1 is a kernel function. □

9.2 Infinitely divisible kernels

Definition 9.10. A kernel K is called $\boxed{\text{INFINITELY DIVISIBLE}}$ if for each $n \geq 1$ there exists a kernel K_n such that $(K_n)^n = K$.

It can happen that K is a kernel function, but $|K|$ is not. We show this by giving an example of a 4×4 matrix such that $A = (a_{i,j}) \geq 0$ but the matrix $(|a_{i,j}|)$ is not positive. Setting $X = \{1, 2, 3, 4\}$ and $K(i, j) = a_{i,j}$ then yields the desired kernel function.

Consider the matrix $A = \begin{pmatrix} 1 & 0 & \frac{1}{\sqrt{2}} & \frac{-1}{\sqrt{2}} \\ 0 & 1 & \frac{1}{\sqrt{2}} & \frac{1}{\sqrt{2}} \\ \frac{1}{\sqrt{2}} & \frac{1}{\sqrt{2}} & 1 & 0 \\ \frac{-1}{\sqrt{2}} & \frac{1}{\sqrt{2}} & 0 & 1 \end{pmatrix}$. Applying the Cholesky algorithm one can show that A is positive semidefinite, and that $(|a_{i,j}|)$ is not positive semidefinite.

From our earlier results it is easily checked that if $K : X \times X \to \mathbb{C}$ is a kernel, then $\overline{K}$ and $|K|^2 = K\overline{K}$ are kernels.

Suppose that ψ is a self-adjoint and conditionally negative definite function and that $K = e^{-\psi}$. We have seen that K is a kernel. In fact, K is infinitely divisible, since $K^t = e^{-t\psi}$. Our next result shows that all infinitely divisible kernels K such that $K(x, y) > 0$ are of this form.

Proposition 9.11. *If $K : X \times X \to \mathbb{C}$ is an infinitely divisible kernel, then $|K|$ is a kernel.*

Proof. Let K_n be a kernel such that $K_n^n = K$. We have $|K| = \left|K_2^2\right| = |K_2|^2$ which is a kernel by our previous result. $\qquad\square$

Theorem 9.12. *Let $K : X \times X \to \mathbb{C}$ be a kernel such that $K(x, y) > 0$ for all x, y. Then K is infinitely divisible if and only if $\psi(x, y) = -\log(K(x, y))$ is conditionally negative definite.*

Proof. Suppose that $\psi(x, y) = -\log(K(x, y))$ is conditionally negative definite. Then since ψ is self-adjoint, $e^{-\psi(x,y)} = K(x, y)$ is a kernel function and setting $K_n = e^{-\psi/n}$ proves that K is infinitely divisible.

Now assume that ψ is defined by $\psi = -\log(K)$ and that K is infinitely divisible. We have

$$e^{-\psi} = K = K_n^n = |K_n|^n.$$

Let m, n be natural numbers and consider

$$e^{-\frac{m}{n}\psi} = \left(e^{-\psi/n}\right)^m = |K_n|^m.$$

Since K_n is an infinitely divisible kernel, $|K_n|$ is a kernel function. It follows that $e^{-t\psi}$ is a kernel function for all rational positive values of t. Since limits of kernel functions are again kernel functions, we see that $e^{-t\psi}$ is a kernel for all $t > 0$. It follows that ψ is conditionally negative definite. $\qquad\square$

9.3 Exercises

Exercise 9.1. Let $\mathcal{H}$ be a Hilbert space. Prove that $\psi(x, y) = \|x - y\|^2$ is a symmetric, normalized, conditionally negative definite function. Give an example to show that it is not a negative definite function.

Exercise 9.2. Let $\|(a_1, \dots, a_n)\|_1 = \sum_{i=1}^n |a_i|$ and let $\|(a_1, \dots, a_n)\|_\infty = \max\{|a_1|, \dots, |a_n|\}$ denote the 1-norm and ∞-norm on $\mathbb{C}^n$. Give examples to show that $\psi_1(x, y) = \|x - y\|_1^2$ and $\psi_\infty(x, y) = \|x - y\|_\infty^2$ are not conditionally negative definite functions on $\mathbb{C}^n$.

Exercise 9.3. Show $\psi(x, y) = \sin^2(x - y)$ is a conditionally negative definite function on $\mathbb{R}$.

Exercise 9.4. If ψ is a self-adjoint conditionally negative definite function and $\psi(x, y) \geq 0$, then ψ^α is a self-adjoint conditionally negative definite function for $0 < \alpha \leq 1$. Prove this result by establishing the fact that $x^\alpha = c_\alpha \int_0^\infty (1 - e^{-xt}) t^{-\alpha/2-1} \, dt$, where c_α is a positive constant. Give a second proof by showing that $x^\alpha = b_\alpha \int_0^\infty \frac{t}{t+x} t^{\alpha-1} dt$ for some positive constant b_α.

Exercise 9.5. Use Schoenberg's theory to prove that there exists a negative definite function ψ such that ψ^α is not negative definite for any $\alpha > 1$.

Exercise 9.6. Here is a more constructive proof of Schoenberg's lemma. Let W be the set of functions on X that have finite support and such that $\sum_{x \in X} f(x) = 0$. Let ψ be conditionally negative definite, $\psi(x, y) \geq 0$ and $\psi(x, x) = 0$ for all $x \in X$. Define a bilinear form on W by $\phi(f, g) = -\frac{1}{2} \sum_{x \in X} \psi(x, y) f(x) g(y)$.

1. Prove that ψ is an inner product. Let $\mathcal{H}$ denote the Hilbert space completion of W in this inner product.
2. Fix a point $p \in X$ and show that the map $\phi(x) = \delta_x - \delta_p$ embeds X into $\mathcal{H}$ in such a way that $\|\phi(x) - \phi(y)\|^2 = \psi(x, y)$.

10

Positive definite functions on groups

10.1 Naimark's theorem

The theory of composition operators is an important tool in the study of unitary representations of groups and gives a very quick proof of a result known as Naimark's dilation theorem. Given a group G with identity e, and a Hilbert space $\mathcal{H}$, we call a homomorphism, $\pi : G \to B(\mathcal{H})$, such that $\pi(e) = I_{\mathcal{H}}$ and $\pi(g^{-1}) = \pi(g)^*$, i.e. such that $\pi(g)$ is unitary, for all $g \in G$, a $\boxed{\text{UNITARY REPRESENTATION OF } G \text{ ON } \mathcal{H}}$. A unitary representation is said to be $\boxed{\text{CYCLIC}}$ if there exists $v_0 \in \mathcal{H}$ such that the linear span of $\pi(G)v_0 := \{\pi(g)v_0 : g \in G\}$ is dense in $\mathcal{H}$.

Definition 10.1. Let G be a group and let $p : G \to \mathbb{C}$ be a function. Then p is called a $\boxed{\text{POSITIVE DEFINITE FUNCTION ON } G}$ provided that for every n and every $g_1, \ldots, g_n \in G$, the matrix $(p(g_i^{-1}g_j))$ is positive semidefinite.

Again, the language varies. Some books call a positive definite function on a group a $\boxed{\text{FUNCTION OF POSITIVE TYPE}}$. This is especially the case when the group is abelian.

Note that saying that p is positive definite is the same as requiring that $K_p : G \times G \to \mathbb{C}$ defined by $K_p(g, h) = p(g^{-1}h)$ is a kernel function. Thus, to every positive definite function on G there is associated an RKHS, $\mathcal{H}(K_p)$. Note that in this space, the kernel function for evaluation at the identity element is $k_e(g) = p(g^{-1})$.

Now fix $g \in G$ and consider the function, $\varphi : G \to G$ defined by $\varphi(h) = g^{-1}h$. We have that $K \circ \varphi(g_1, g_2) = K(g^{-1}g_1, g^{-1}g_2) = p((g^{-1}g_1)^{-1}(g^{-1}g_2)) = K(g_1, g_2)$. Thus, by Theorem 5.10 there is a well-defined contractive linear map, $U_g : \mathcal{H}(K_p) \to \mathcal{H}(K_p)$ with $(U_g f)(h) = f(g^{-1}h)$ for any $f \in \mathcal{H}(K_p)$. Now $(U_{g_1} \circ U_{g_2})(f)(h) = (U_{g_2}f)(g_1^{-1}h) = f(g_2^{-1}g_1^{-1}h) = (U_{g_1 g_2}f)(h)$, and so the map $\pi : G \to B(\mathcal{H}(K_p))$, is a homomorphism. Since $U_{g^{-1}} \circ U_g = I_{\mathcal{H}(K_p)}$, and both of these maps are contractions, it follows that they must both be invertible isometries and, hence, unitaries.

Thus, to every positive definite function on p on G, we have associated a unitary representation, $\pi : G \to B(\mathcal{H}(K_p))$, by setting $\pi(g) = U_g$. This construction gives an immediate proof of a theorem of Naimark.

Theorem 10.2 (Naimark's dilation theorem). *Let G be a group and let p : $G \to \mathbb{C}$ be a positive definite function. Then there exists a Hilbert space $\mathcal{H}$, a unitary representation $\pi : G \to B(\mathcal{H})$ and a vector $v \in \mathcal{H}$, such that $p(g) = \langle \pi(g)v, v \rangle$. Moreover, any function of this form is positive definite.*

Proof. Let $\mathcal{H} = \mathcal{H}(K_p)$, let $\pi(g) = U_g$ and let $v = k_e$. We have that $\langle \pi(g)k_e, k_e \rangle = (\pi(g)k_e)(e) = k_e(g^{-1}) = p(g)$.

Finally, if $f(g) = \langle \pi(g)v, v \rangle$, and we pick $\{g_1, \ldots, g_n\} \subseteq G$ and scalars $\alpha_1, \ldots, \alpha_n \in \mathbb{C}$, then we have that

$$\sum_{i,j=1}^{n} \overline{\alpha_i}\alpha_j f(g_i^{-1}g_j) = \langle w, w \rangle$$

where $w = \sum_{j=1}^{n} \alpha_j \pi(g_j)v$ and so p is a positive definite function. $\square$

Given a positive definite function p on a group, the representation that we get by considering composition operators on $\mathcal{H}(K_p)$ is also cyclic, with cyclic vector k_e. To see this, note that $(U_g k_e)(h) = k_e(g^{-1}h) = K(g^{-1}h, e) = p(h^{-1}g) = K(h, g) = k_g(h)$. Thus, $U_g k_e = k_g$ and, hence, the span of $\pi(G)k_e$ is equal to the span of $\{k_g : g \in G\}$, which is always dense in the RKHS.

Conversely, assume that we have a unitary representation γ of G on some Hilbert space $\mathcal{H}$, which has a cyclic vector v_0, and we define $p(g) = \langle \gamma(g)v_0, v_0 \rangle$. By Naimark's theorem, we know that p is positive definite.

Now consider the Hilbert space, $\mathcal{H}(K_p)$ and unitary representation π of G. We claim that there is a Hilbert space isomorphism, $W : \mathcal{H}(K_p) \to \mathcal{H}$, such that $Wk_e = v_0$ and $W\pi(g) = \gamma(g)W$, for all $g \in G$.

To define W we set $Wk_g = \gamma(g)v_0$, and extend linearly. Note that

$$\left\| \sum_i \alpha_i k_{g_i} \right\|^2 = \sum_{i,j} \alpha_i \overline{\alpha_j} k_{g_i}(g_j) = \sum_{i,j} \alpha_i \overline{\alpha_j} p(g_j^{-1}g_i)$$

$$= \sum_{i,j} \alpha_i \overline{\alpha_j} \langle \gamma(g_j^{-1}g_i)v_0, v_0 \rangle = \left\| \sum_i \alpha_i \gamma(g_i)v_0 \right\|^2.$$

This equality shows that W is well-defined and an isometry. Thus, W can be extended by continuity to an isometry from all of $\mathcal{H}(K_p)$ onto $\mathcal{H}$. Finally,

$$W\pi(g)k_{g_1} = W\pi(g)\pi(g_1)k_e = W\pi(gg_1)k_e =$$
$$Wk_{gg_1} = \gamma(gg_1)v_0 = \gamma(g)\gamma(g_1)v_0 = \gamma(g)Wk_{g_1}$$

and since these vectors span the space, $W\pi(g) = \gamma(g)W$.

Thus, the representation γ is unitarily equivalent to the representation π, via a map that carries the cyclic vector v_0 to the vector k_e.

These calculations show that if one requires the vector v appearing in the dilation of a positive definite function in Naimark's dilation theorem to be cyclic, then up to a unitary equivalence, we are in the situation where $\mathcal{H} = \mathcal{H}(K_p)$, $\pi(g) = U_g$ and $v = k_e$.

10.2 Bochner's theorem

Bochner's theorem gives a complete characterization of the positive definite functions, i.e. functions of positive type, on an abelian group. A full proof of Bochner's theorem is outside the scope of our book, so we will only present some of the details. The interested reader should look at [9].

In the previous section we saw that there is a one-to-one correspondence between positive definite functions on a group G and cyclic unitary representations of the group G. Bochner's theorem tells us that for abelian groups every continuous positive type function is the Fourier transform of a positive measure.

Definition 10.3. Let G be an abelian group. A homomorphism $\phi : G \to \mathbb{T}$ is called a character. The set of all characters is called the dual group and is denoted $\hat{G}$.

The dual group actually is a group. Note that the pointwise product of two characters is again a character and that the pointwise inverse of a character is again a character. The function that is constantly 1 is a character and this is the identity element of $\hat{G}$.

Since a character is a 1-dimensional unitary representation we see that ϕ is of positive type.

If G is a locally compact abelian group, then $\hat{G}$ is a locally compact group under the topology of uniform convergence on compact sets. That is, $\phi_t \to \phi$ if and only if for every compact set $K \subseteq G$ we have $\sup_{g \in K} |\phi_t(g) - \phi(g)| \to 0$. Every locally compact abelian group carries a unique translation invariant measure, known as Haar measure. The Fourier transform of a function $f \in L^1(G)$ is given by

$$\hat{f}(p) = \int_G f(g)\overline{p(g)}\, dg,$$

where p is a character and dg is Haar measure on G. Since the function $p : G \to \mathbb{C}$ is bounded, we see that for a finite measure μ on G that

$$\hat{\mu}(p) = \int_G \overline{p(g)}\, d\mu(g)$$

is well-defined. We call $\hat{\mu}$ the Fourier transform of μ.

Theorem 10.4 (Bochner's theorem). *If μ is a finite nonnegative Borel measure on a locally compact abelian group G, then $\hat{\mu}$ is a continuous function of positive type on $\hat{G}$. Conversely, if f is a continuous function of positive type on $\hat{G}$ then there is a finite nonnegative Borel measure μ on G such that $f = \hat{\mu}$.*

Proof. The first direction of Bochner's theorem is relatively straightforward and we prove it, while the converse requires more measure theory than we are prepared to develop.

Let $\alpha_1, \ldots, \alpha_n \in \mathbb{C}$ and let $p_1, \ldots, p_n \in \hat{G}$. We have,

$$\sum_{i,j=1}^n \overline{\alpha_i}\alpha_j \hat{\mu}(p_j^{-1}p_i) = \int_G \sum_{i,j=1}^n \overline{\alpha_i}\alpha_j \overline{p_j^{-1}p_i(g)}\, d\mu(g)$$

$$= \int_G \left| \sum_{i=1}^n \alpha_i p_i(g) \right|^2 d\mu(g) \geq 0$$

$\square$

10.3 Negative definite functions and cocycles

In Theorem 10.2 we showed that there is a correspondence between positive definite functions on a group and representations of the group. We now show that there is a correspondence between conditionally negative definite functions, group *cocycles* and affine representations of groups. This allows us to relate conditionally negative definite functions and affine actions of groups.

Definition 10.5. An $\boxed{\text{AFFINE TRANSFORMATION}}$ of a Hilbert space $\mathcal{H}$ is a map of the form $f(x) = Tx + v$, where T is a bounded operator and v is a vector in $\mathcal{H}$.

Thus, an affine transformation is the composition of a linear transformation followed by translation by a fixed vector. It is not hard to see that an affine transformation is invertible if and only if the linear transformation T is invertible, and that in this case the inverse is also an affine transformation. In fact, $f^{-1}(x) = T^{-1}(x) - T^{-1}(v)$.

If T is a unitary, then the corresponding affine action of a Hilbert space is an onto isometry of H, viewed as a metric space.

We shall use the fact that, conversely, every onto isometry of a real Hilbert space is of this form. That is, if $f : \mathcal{H} \to \mathcal{H}$ is an onto isometry, then $f(x) = Ux + v$ where U is a unitary and v is a vector in $\mathcal{H}$. Furthermore, the unitary U and the vector v are uniquely determined by f.

A group is said to *act by affine isometries on* $\mathcal{H}$ if and only if there is a homomorphism from G into the group of affine isometries. If α is a such an action, then there exists a unitary $U(g)$ and a vector $b(g)$ such that $\alpha_g(x) = U(g)x + b(g)$.

Definition 10.6. Let $g \to U_g$ be a unitary representation of a group G on a Hilbert space $\mathcal{H}$. A map $b : G \to \mathcal{H}$, written as $g \to b_g$, is called a $\boxed{\text{COCYCLE}}$ for the representation if and only if it satisfies the $\boxed{\text{COCYCLE LAW}}$: $b_{gh} = U_g b_h + b_g$.

Theorem 10.7. *Let α be an affine isometric action of G on $\mathcal{H}$. Then the map $g \mapsto U_g$ is a group representation and the map $g \mapsto b_g$ is a cocycle for the representation.*

Proof. Since α is an action we have $\alpha_{gh} = \alpha_g \circ \alpha_h$. This gives $U_{gh}(x) + b_{gh} = U_g(U_h(x) + b_h) + b_g = U_g U_h(x) + U_g b_h + b_g$. It follows from the uniqueness of the unitary and the vector that $U_{gh} = U_g U_h$ and $b_{gh} = U_g b_h + b_g$. $\qquad\square$

The converse of the above result is also true.

Theorem 10.8. *Let $\pi : G \to B(\mathcal{H})$ be a unitary representation, set $U_g = \pi(g)$, let $b : G \to \mathcal{H}$ be a cocycle, and define $\alpha_g(x) = U_g(x) + b_g$. Then α is an affine action of G on $\mathcal{H}$.*

We now make the connection between negative definite functions and cocycles.

Definition 10.9. Let G be a group. A function $\gamma : G \to \mathbb{C}$ is of $\boxed{\text{NEGATIVE TYPE}}$ if and only if the function $\psi : G \times G \to \mathbb{C}$ defined by $\psi(g,h) = \gamma(g^{-1}h)$ is conditionally negative definite. We say that γ is symmetric if $\gamma(g^{-1}) = \gamma(g)$ and self-adjoint if $\gamma(g^{-1}) = \overline{\gamma(g)}$.

Note that the definitions are made so that γ is symmetric if and only if ψ is symmetric and γ is self-adjoint if and only if ψ is self-adjoint.

First, we make some simple observations. Let $b : G \to \mathcal{H}$ be a cocycle for some representation U_g. The cocycle law implies certain conditions on b. We have, $b_e = b_{e^2} = U_e b_e + b_e = 2b_e$ and so $b_e = 0$. We also have $0 = b_e = b_{gg^{-1}} = U_g b_{h^{-1}} + b_g$ and so $b_g = -U_g b_{g^{-1}}$. Hence, $\|b_g\| = \|b_{g^{-1}}\|$.

Theorem 10.10. *Let G be a group. If $b : G \to \mathcal{H}$ is a cocycle with respect to a unitary representation U_g, then $\gamma(g) = \|b_g\|^2$ is symmetric, of negative type, and $\gamma(e) = 0$. Conversely if γ is a symmetric function of negative type on G with $\gamma(e) = 0$, then there exists a unitary representation U_g on $\mathcal{H}$ and a cocycle b, such that $\gamma(g) = \|b_g\|^2$.*

Proof. Since b is an embedding of G into a Hilbert space, the map b induces a conditionally negative definite function by $\psi(g,h) = \|b_g - b_h\|^2$. Applying the cocycle identity we get that $-b_h = U_h b_{h^{-1}}$, which gives $\psi(g,h) = \|b_g + U_h b_{h^{-1}}\|^2$. Since U_h is an isometry we get $\psi(g,h) = \|U_{h^{-1}} b_g + b_{h^{-1}}\|^2 = \|b_{h^{-1}g}\|^2 = \|b_{g^{-1}h}\|^2$. It follows that the function $\gamma(g) = \|b_g\|^2$ is a function of negative type. Also, since $b_e = 0$, $\gamma(e) = 0$.

Now assume that $\gamma : G \to \mathbb{C}$ is symmetric of negative type and $\gamma(e) = 0$. Then $\psi(g,h) = \gamma(g^{-1}h)$ is conditionally negative definite and $\psi(g,g) = \gamma(e) = 0$, so that ψ is normalized and $\psi(h,g) = \psi(h,g)$. Applying Schoenberg's lemma 9.5, we see that there is a map $b : G \to \mathcal{H}$ such that $\gamma(h^{-1}g) = \gamma(g^{-1}h) = \|b_g - b_h\|^2$ for all $g,h \in G$. Clearly, if we translate b_g by any fixed vector then it still satisfies this equation. Thus, after translating by b_e we may assume that $b_e = 0$. Hence, $\gamma(g) = \|b_g - b_e\|^2 = \|b_g\|^2$.

The problem now is to find a representation π of the group G on $\mathcal{H}$ such that (π, b) satisfy the cocycle law.

Let $K(g,h) = -\psi(g,h) + \psi(g,e) + \psi(e,h) - \psi(e,e) = -\gamma(g^{-1}h) + \gamma(g^{-1}) + \gamma(h)$ denote the kernel function associated to ψ by Proposition 9.3. Note that $K(e,e) = \gamma(e) = 0$ so that in $\mathcal{H}(K)$ we have that $k_e = 0$.

Also, note that in the proof of Schoenberg's lemma, for the space $\mathcal{H}$ we used $\mathcal{H}(K)$.

Define a map on the span of the kernel functions by $U_g k_t = k_{gt} - k_g$, for $g,t \in G$. We claim that U_g is a unitary and that $U_{gh} = U_g U_h$. Let $f = \sum_{j=1}^n \alpha_j k_{t_j}$ be an element of $\mathcal{H}(K)$. We have

$$\|U_g f\|^2 = \sum_{i,j=1}^{n} \overline{\alpha_i}\alpha_j \left\langle U_g k_{t_j}, U_h k_{t_i} \right\rangle$$

$$= \sum_{i,j=1}^{n} \overline{\alpha_i}\alpha_j \left\langle k_{gt_j} - k_g, k_{gt_i} - k_g \right\rangle.$$

Using that γ is symmetric, we have $K(s,t) = -\gamma(t^{-1}s) + \gamma(s) + \gamma(t^{-1})$. We have

$$\begin{aligned}
\left\langle k_{gt} - k_g, k_{gs} - k_g \right\rangle &= K(gs, gt) + K(g, g) - K(g, gt) - K(gs, g)\\
&= -\gamma((gt)^{-1}gs) + \gamma(gs) + \gamma((gt)^{-1})\\
&\quad - \gamma(g^{-1}g) + \gamma(g) + \gamma(g^{-1})\\
&\quad + \gamma((gt)^{-1}g) - \gamma(g) - \gamma((gt)^{-1})\\
&\quad + \gamma(g^{-1}gs) - \gamma(gs) - \gamma(g^{-1})\\
&= -\gamma(t^{-1}s) + \gamma(s) + \gamma(t^{-1}) = K(s,t).
\end{aligned}$$

From this we see that $\|Uf\| = \|f\|$. Therefore, U extends uniquely to an isometry on $\mathcal{H}$. We have $U_e k_t = k_t - k_e = k_t$, since k_e is the zero vector, hence $U_e = I$. Finally,

$$\begin{aligned}
U_{gh} k_t &= k_{ght} - k_{gh} = k_{g(ht)} - k_g - (k_{gh} - k_g)\\
&= U_g(k_{ht}) - U_g(k_h) = U_g(k_{ht} - k_h) = U_g(U_h(k_t)).
\end{aligned}$$

It follows that $I = U_g U_{g^{-1}}$ and hence, U_g is an invertible map. Therefore we have a unitary representation of G on $\mathcal{H}$.

The cocycle condition is now easy to check: we have $b(gh) = k_{gh} = U_g k_h + k_g = U_g b(h) + b(g)$. $\qquad\square$

10.4 Exercises

Exercise 10.1. Let $\gamma : G \to \mathbb{C}$ and set $\psi(g, h) = \gamma(g^{-1}h)$. Prove that γ is symmetric if and only if ψ is symmetric and that γ is self-adjoint if and only if ψ is self-adjoint.

Exercise 10.2. Prove that the function $p(x) = cos(x)$ is a positive definite function on the group $(\mathbb{R}, +)$. Show that $\mathcal{H}(K_p)$, is two-dimensional and explicitly describe the unitary representation of $\mathbb{R}$ on this space.

Exercise 10.3. Let G be an abelian group, and let δ_g be the point mass concentrated at $g \in G$. Show that the Fourier transform of δ_g is a character.

Exercise 10.4. Give an example of a function $p : (\mathbb{R}, +) \to \mathbb{C}$ that is of positive type and is not the Fourier transform of a measure. Note that p cannot be continuous.

Exercise 10.5. Let $p(z) = \sum_{k=0}^{n} a_k z^k$, with $a_k \geq 0, k = 0, \ldots, n$. Show that p is a positive definite function on the circle group, $(\mathbb{T}, \cdot)$ and that $\mathcal{H}(K_p)$ is $n + 1$-dimensional. Explicitly describe the unitary representation.

Exercise 10.6. Let $p(n) = \begin{cases} 1 & n \text{ even} \\ 0 & n \text{ odd} \end{cases}$. Prove that p is a positive definite function on $(\mathbb{Z}, +)$, find $\mathcal{H}(K_p)$ and the corresponding unitary representation.

Exercise 10.7. Compute the Fourier transform of the characteristic function of the interval $[-A, A]$.

11
Applications of RKHS to integral operators

In this chapter we show how the theory of reproducing kernel Hilbert spaces is related to the theory of integral operators. In particular, we will give a proof of Mercer's theorem that uses RKHS techniques in a crucial manner and we will study the relationship between the Volterra integral operator and the RKHS induced by the "min" kernel.

11.1 Integral operators

Given a measure space (Y, μ) and a function $S : X \times Y \to \mathbb{C}$ with the property that for each $x \in X$ the function $S(x, y)$ is a square-integrable function of y, then for each square-integrable function g one obtains a function $f : X \to \mathbb{C}$ by setting

$$f(x) = \int_X S(x, y)g(y)d\mu(y).$$

If $g_1 = g$ a.e. μ, then

$$\int_X S(x, y)g(y)d\mu(y) = \int_X S(x, y)g_1(y)d\mu(y),$$

and so the function $f(x)$ really only depends on the equivalence class of g. Thus, setting $T_S g = f$ gives a well-defined map from $L^2(Y, \mu)$ into a space of functions on X. It is easy to see that T_S is a linear map and so the range of T_S is a vector space of functions on X.

Such an operator T_S is called an $\boxed{\text{INTEGRAL OPERATOR}}$ and the function S is called the $\boxed{\text{INTEGRAL KERNEL}}$ or $\boxed{\text{SYMBOL}}$ of the integral operator. We shall prefer the word "symbol" to minimize confusion with our meaning of the word kernel, but the phrase "integral kernel" is widely used. To further confuse notations between the two fields, a widely studied family of integral operators are those for which $X = Y$ and the symbol S is a positive semidefinite function, i.e. a kernel function in our sense!

Integral operators arise quite often in the study of differential equations as the inverses of differential operators.

To develop an intuition for the behavior of such operators it is helpful to look first at the case where $X = \{1, 2, \ldots, m\}$, $Y = \{1, \ldots, n\}$ are finite discrete spaces and μ is counting measure, so that $\mu(\{j\}) = 1$, for all j. In this case we have that

$$(T_S g)(i) = \int_X S(i, j)g(j)d\mu(j) = \sum_{j=1}^{n} S(i, j)g(j),$$

so that if we identify functions on X and Y with vectors in $\mathbb{C}^m$ and $\mathbb{C}^n$, respectively, then the integral operator is just multiplication by the $m \times n$ matrix $(S(i, j))$. Thus, if one thinks of integration as a "continuous sum", then integral operators are just "continuous" versions of matrix multiplication.

The overlap of notations between integral operators and the field of reproducing kernel Hilbert spaces is not a coincidence and does cause one to wonder:

- Is the range of an integral operator an RKHS?
- When S is a positive semidefinite function, is there a relationship between the operator T_S and the RKHS $\mathcal{H}(S)$?
- Does the theory of RKHS have any useful applications in this setting?

In this chapter we will address these questions and use the theory of RKHS to give proofs of some of the results in the area of integral operators. We begin by analyzing the range of an integral operator.

11.2 The range space of an integral operator

In this section we prove that the range of an integral operator is an RKHS and give a formula for the reproducing kernel of this space in terms of the symbol of the integral operator. To do this it will be necessary to introduce an extra operation on symbols.

Definition 11.1. Given sets X and Z, a measure space (Y, μ) and functions $S_1 : X \times Y \to \mathbb{C}$, $S_2 : Y \times Z \to \mathbb{C}$, both of which are square-integrable functions in the y variable for each fixed x and z, we define their $\boxed{\text{BOX PRODUCT}}$ to be the function $S_1 \square S_2 : X \times Z \to \mathbb{C}$ given by

$$S_1 \square S_2(x, z) = \int_Y S_1(x, y) S_2(y, z) d\mu(y).$$

We also define the $\boxed{\text{ADJOINT OF A FUNCTION}}$ to be the function $S_1^* : Y \times X \to \mathbb{C}$ given by $S_1^*(y, x) = \overline{S_1(x, y)}$.

The motivation for both of these definitions comes from thinking of integral operators as "continuous matrix products", in which case the box product is just the analogue of the matrix product and S_1^* is the analogue of the adjoint of a matrix.

We begin with a preliminary result that should be recognizable from matrix theory.

Proposition 11.2. *Let X be a set, let (Y, μ) be a measure space and let S : $X \times Y \to \mathbb{C}$ be square integrable in the y variable. Then the function $S \square S^*$: $X \times X \to \mathbb{C}$ is a kernel function.*

Proof. Set $K = S \square S^*$, fix points $x_1, \ldots, x_n \in X$ and scalars $\lambda_1, \ldots, \lambda_n \in \mathbb{C}$. Let $g(y) = \sum_{i=1}^n \overline{\lambda_i} S(x_i, y) \in L^2(Y, \mu)$ and compute

$$\sum_{i,j=1}^n K(x_i, x_j) \overline{\lambda_i} \lambda_j = \int_Y \sum_{i,j=1}^n \overline{\lambda_i} S(x_i, y) \overline{S(x_j, y)} \lambda_j d\mu(y) = \int_Y |g(y)|^2 d\mu(y).$$

$\square$

Before we continue with our discussion we need to introduce the $\boxed{\text{RANGE NORM}}$ of a Hilbert space operator [12]. Suppose that $\mathcal{H}$ and $\mathcal{K}$ are Hilbert spaces and that $T : \mathcal{H} \to \mathcal{K}$ is a bounded operator. Given a vector $h \in \mathcal{R}(T)$ we define $\|h\|_{\mathcal{R}(T)} = \inf\{\|g\| : Tg = h\}$. It is not hard to show that (Exercise 11.1) equipped with this norm the vector space $\mathcal{R}(T)$ is a Hilbert space. This is somewhat surprising, since $\mathcal{R}(T)$ is not necessarily a closed subspace of $\mathcal{K}$. When $\mathcal{R}(T)$ is equipped with this range norm, the operator $T|_{\mathcal{K}(T)^\perp} : \mathcal{K}(T)^\perp \to \mathcal{R}(T)$ is a surjective isometry.

We can now state the main result of this section.

Theorem 11.3. *Let X be a set, let (Y, μ) be a measure space, let $S : X \times Y \to \mathbb{C}$ be square integrable in the y variable and let $K : X \times X \to \mathbb{C}$ be the kernel function given by $K(x, z) = (S \square S^*)(x, z)$. Then $\{T_S g : g \in L^2(Y, \mu)\} = \mathcal{H}(K)$ and for $f \in \mathcal{H}(K)$ we have that*

$$\|f\|_{\mathcal{H}(K)} = \inf\{\|g\|_2 : f = T_S g\}.$$

Proof. Let $\mathcal{R} = \{T_S g : g \in L^2(Y, \mu)\}$ and for $f \in \mathcal{R}$ set

$$\|f\|_{\mathcal{R}} = \inf\{\|g\|_2 : f = T_S g\},$$

i.e. we let $\mathcal{R}$ be the range space in the sense of Sarason. Since $\mathcal{R}$ is a concrete space of functions on X, the proof will be complete if we can show that $\mathcal{R}$ is an RKHS and that K is the corresponding kernel function.

To see that the evaluation functionals are bounded, we compute

$$|f(x)| = |\int_Y S(x, y) g(y) d\mu(y)| \leq$$

$$\left(\int_Y |S(x, y)|^2 d\mu(y)\right)^{1/2} \left(\int_Y |g(y)|^2 d\mu(y)\right)^{1/2}.$$

Since this inequality holds for any g satisfying $f = T_S g$, we have that

$$|f(x)| \leq \|h_x\|_2 \|f\|_{\mathcal{R}},$$

where $h_x(y) = S(x, y)$.

Recall that when we endow a range space with the Sarason norm and let $\mathcal{V} = \mathcal{K}(T_S)^{\perp}$, then $T_S : \mathcal{V} \to \mathcal{R}$ is an onto isometry. Hence, for any two functions $f_1, f_2 \in \mathcal{R}$ we have that

$$\langle f_1, f_2 \rangle_{\mathcal{R}} = \langle g_1, g_2 \rangle_{L^2(Y)},$$

where $g_1, g_2 \in \mathcal{V}$ are the unique functions satisfying $T_S g_i = f_i$.

Note that $w(y) \in \mathcal{K}(T_S)$ if and only if $\int_Y h_x(y) w(y) d\mu(y) = 0$ for all x; that is, if and only if $w \perp \overline{h_x}$ for all x. Hence, $\mathcal{V} = \mathcal{K}(T_S)^{\perp}$ is the closed linear span of $\{\overline{h_x} : x \in X\}$.

In particular, $g_t = \overline{h_t} \in \mathcal{V}$ and if we set $k_t = T_S(g_t)$, then for any $f = T_S g$ with $g \in \mathcal{V}$, we have that

$$\langle f, k_t \rangle_{\mathcal{R}} = \langle g, g_t \rangle_{L^2(Y)} = \int_Y \overline{g_t(y)} g(y) d\mu(y) =$$

$$\int_Y S(t, y) g(y) d\mu(y) = f(t).$$

Thus, $\mathcal{R}$ is an RKHS and k_t is the kernel function for the point t. Finally,

$$K(x, y) = k_t(x) = (T_S g_t)(x) = \int_Y S(x, y) g_t(y) d\mu(y) =$$

$$\int_Y S(x, y) \overline{S(t, y)} d\mu(y) = \int_Y S(x, y) S^*(y, t) d\mu(y) = (S \square S^*)(x, t)$$

and the proof is complete. $\qquad\qquad\qquad\qquad\qquad\qquad\qquad\qquad\square$

Corollary 11.4. *Let* X *be a set, let* (Y, μ) *be a measure space, let* $S : X \times Y \to \mathbb{C}$ *be square integrable in the* y *variable, let* $K : X \times X \to \mathbb{C}$ *be the kernel function given by* $K(x, z) = (S \square S^*)(x, z)$ *and set* $g_x(\cdot) = \overline{S(x, \cdot)} \in L^2(Y, \mu)$. *Then there is an isometry,* $U : \mathcal{H}(K) \to L^2(Y, \mu)$ *with* $U(k_x) = g_x$.

The following gives a nice illustration of this theorem.

11.2.1 The Volterra integral operator and the min kernel

Let $X = Y = [0, +\infty)$ and let $\mu = m$ denote the standard Lebesgue measure. The $\boxed{\text{VOLTERRA INTEGRAL OPERATOR}}$ is the integral operator V with domain $L^2([0, +\infty), m)$ given by

$$(Vg)(x) = \int_0^x g(y) dm(y).$$

Note that if we define $S : X \times Y \to \mathbb{C}$ by

$$S(x, y) = \begin{cases} 1, & 0 \le y \le x \\ 0, & x < y \end{cases}$$

then V is the integral operator with symbol S.

From standard measure theory, we know that a function f is of the form $f(x) = (Vg)(x)$ if and only if f is absolutely continuous, $f'(x) = g(x)$ almost everywhere with $f' \in L^2([0, +\infty), m)$ and $f(0) = 0$.

Computing

$$K(x, t) = S \square S^*(x, t) = \int_0^{+\infty} S(x, y) S(t, y) dy = \min\{x, t\}.$$

Earlier, in Section 2.3.2 we had studied the RKHS $\mathcal{H}(K)$ for this kernel. Now, through the theory of integral operators we have a complete understanding of the functions that are in this space and the characterization of the norm. Namely, for $K(x, t) = \min\{x, t\}$, we have

$\mathcal{H}(K) = \mathcal{R}(V)$

$\quad = \{f : [0, +\infty) \to \mathbb{C} | f(0) = 0, f \text{ absolutely continuous}, f' \in L^2\}$

and $\|f\|_{\mathcal{H}(K)} = \|f'\|_2$.

11.3 Mercer kernels and integral operators

In this section we prove Mercer's theorem, which is a classical result in functional analysis. Recently, the result has been used in the probability and machine learning community.

The standard approach to Mercer's theorem relies on the theory of compact operators and the connection between Hilbert–Schmidt operators and integral operators whose symbols are square-integrable on the product domain. Here we present a somewhat more elementary proof that uses techniques from the theory of reproducing kernel Hilbert spaces and the Arzelà–Ascoli theorem.

For the remainder of this section we will assume that X is a compact subset of $\mathbb{R}^d$ and that μ is a finite Borel measure on X. We will write $d\mu(t)$ as dt.

Definition 11.5. Let X be a compact subset of $\mathbb{R}^d$. A function $K : X \times X \to \mathbb{C}$ is called a $\boxed{\text{MERCER KERNEL}}$ if it is a continuous, positive semidefinite function, i.e. a continuous kernel function.

We begin by studying the RKHS $\mathcal{H}(K)$ when K is a Mercer kernel. First, note that since K is continuous, by Theorem 2.17 every function in $\mathcal{H}(K)$ is continuous.

Proposition 11.6. *Let K be a Mercer kernel on X and let μ be a finite Borel measure on X. Then there are constants $B, C > 0$ such that for every $f \in \mathcal{H}(K)$, $B\|f\|_2 \leq \|f\|_\infty \leq C\|f\|_{\mathcal{H}(K)}$.*

Proof. The first inequality is trivial if $\mu(X) = 0$. Otherwise set $B = \mu(X)^{-1/2}$, then

$$\|f\|_2^2 = \int_X |f(t)|^2 d\mu(t) \leq \|f\|_\infty^2 \mu(X)$$

and the first inequality follows.

For the second inequality, since K is continuous and X is compact, we may let $C = \sup\{K(x, x)^{1/2} : x \in X\}$, then for any $f \in \mathcal{H}$, by Cauchy–Schwarz,

$$|f(x)| = |\langle f, k_x \rangle| \leq \|f\|_{\mathcal{H}(K)} K(x, x)^{1/2}$$

and the second inequality follows. $\qquad\qquad\qquad\Box$

Proposition 11.7. *Let K be a Mercer kernel, then $\mathcal{H}(K)$ is separable.*

Proof. Since $X \times X$ is compact, K is uniformly continuous. Hence, given $n \in \mathbb{N}$, we may choose finitely many points $F_n = \{y_1, \ldots, y_m\} \subseteq X$, such that for each $y \in X$ there exists i such that $|K(x, y) - K(x, y_i)| < 1/n$. From this it follows that

$$\|k_y - k_{y_i}\|^2 = K(y, y) - K(y, y_i) - K(y_i, y) + K(y_i, y_i) < 2/n.$$

Hence, the span of the functions $\{k_y : y \in \cup_{n \in \mathbb{N}} F_n\}$ is dense in $\mathcal{H}(K)$ and it follows that $\mathcal{H}(K)$ is separable. $\qquad\square$

Because $\mathcal{H}(K)$ is separable, it has a countable orthonormal basis, $\{e_n(x) : n \in \mathbb{N}\}$ and we will have that

$$K(x, y) = \sum_{n=1}^{+\infty} e_n(x)\overline{e_n(y)}.$$

In general, the convergence of such a series is pointwise. Our next lemma shows that in the case of a Mercer kernel the convergence of such a series is absolute and uniform on $X \times X$.

Proposition 11.8. *Let K be a Mercer kernel on X and let $\{e_n(x) : n \in \mathbb{N}\}$ be an orthonormal basis for $\mathcal{H}(K)$ with $K(x, y) = \sum_{n=1}^{\infty} e_n(x)\overline{e_n(y)}$. Then the series converges absolutely and uniformly on X to K.*

Proof. We have $K(x, x) = \sum_{n=1}^{\infty} |e_n(x)|^2$. Since K is continuous we know that $|K(x, y)| \leq C$ for some fixed constant C.

Let $g_n(x) = \sum_{i=n+1}^{\infty} |e_i(x)|^2$. Note that $g_{n+1}(x) \leq g_n(x)$ for all $x \in X$ and that for each $x \in X$, $g_n(x) \to 0$ as $n \to \infty$. We claim that $\{g_n\}$ converges uniformly to 0. To see this, let $\epsilon > 0$. Then, for each $x \in X$ there exists N_x such that $g_n(x) < \epsilon/2$ if $n \geq N_x$. Since g_n is continuous at the point x we see that there is a neighborhood U_x of x such that $|g_n(t)| < \epsilon$ for $t \in U_x$ and $n \geq N_x$. Since X is compact there exist finitely many points $x_1, \ldots, x_n \in X$ such that $X = U_{x_1} \cup \cdots \cup U_{x_n}$. If we let $N = \max\{N_{x_1}, \ldots, N_{x_n}\}$, then we see that for $n \geq N$, we have $|g_n(x)| < \epsilon$ for all $x \in X$.

Now, applying the Cauchy–Schwarz inequality, for $n \geq N$ we have

$$\left| \sum_{i=n+1}^{\infty} e_i(x)\overline{e_i(y)} \right| \leq \sum_{i=n+1}^{\infty} |e_i(x)\overline{e_i(y)}| \leq \sqrt{g_n(x)g_n(y)} < \epsilon$$

for all $x, y \in X$. Hence, the series converges uniformly and absolutely on X to K. $\qquad\square$

We now turn our attention to integral operators defined by K. For this we will want a finite Borel measure μ on X. Recall that a point $x \in X$ is said to be in the $\boxed{\text{SUPPORT OF } \mu}$ provided $\mu(U) \neq 0$ for any open set U containing x.

The support of μ is a closed subset $S \subseteq X$ with the property that $\mu(X \setminus S) = 0$. Given a Mercer kernel on X it would also be a Mercer kernel on S, which is where μ truly "lives", and since $\mu(X \setminus S) = 0$, there is little that can be said about behaviors of integral operators on this set. For this reason we shall often assume that the support of μ is all of X.

Proposition 11.9. *Let $K : X \times X \to \mathbb{C}$ be a Mercer kernel with μ a finite Borel measure on X. Then the integral operator T_K is a bounded map from $L^2(X, \mu)$ into itself with*

$$\|T_K\| \leq \left(\int_X \int_X |K(x, y)|^2 d\mu(x) d\mu(y) \right)^{1/2}$$

and every function in the range of T_K is continuous.

Proof. Set $K^{(2)}(x, z) = K \square K^*(x, z) = K \square K(x, z)$. Let $(x_n, z_n) \to (x, z)$ in $X \times X$. Since the Mercer kernel K is continuous and X is compact, it follows that K is uniformly continuous on $X \times X$. Hence, $K(x_n, y) K(y, z_n) \to K(x, y) K(y, z)$ uniformly in y. By the bounded convergence theorem it follows that

$$K^{(2)}(x, z) = \lim_n \int_X K(x_n, y) K(y, z_n) d\mu(y) = \lim_n K^{(2)}(x_n, y_n).$$

Hence, the function $K^{(2)} = K \square K^*$ is also continuous.

By Theorem 2.17, we will have that every function in $\mathcal{H}(K^{(2)})$ will be continuous on X. But by Theorem 11.3, this RKHS is the range of the operator T_K.

To see that the operator $T_K : L^2(X, \mu) \to L^2(X, \mu)$ is bounded, we let $g \in L^2(X, \mu)$ and compute

$$\|T_K g\|_2^2 = \int_X |T_K g(x)|^2 d\mu(x) \leq \int_X \int_X |K(x, y) g(y)|^2 d\mu(y) d\mu(x) \leq$$

$$\int_X \left(\int_X |K(x, y)|^2 d\mu(y) \right)^2 \left(\int_X |g(y)|^2 d\mu(y) \right)^2 d\mu(x) =$$

$$\left(\int_X \int_X |K(x, y)|^2 \right)^2 \|g\|_2^2.$$

So in particular, $\|T_K\| \leq \|K\|_2$, where K is regarded as a function in $L^2(X \times X, \mu \times \mu)$. $\square$

Proposition 11.10. *Let K be a Mercer kernel on X and let μ be a finite Borel measure. Then the operator $T_K : L^2(X, \mu) \to L^2(X, \mu)$ is positive.*

Proof. Let $g \in L^2(X, \mu)$ and set $K_N(x, y) = \sum_{n=1}^{N} e_n(x)\overline{e_n(y)}$, where $\{e_n(x) : n \in \mathbb{N}\}$ is an orthonormal basis for $\mathcal{H}(K)$. By Proposition 11.8 $K(x, y) - K_N(x, y)$ goes uniformly to 0 on $X \times X$.

If we choose N large enough that $|K(x, y) - K_N(x, y)| < \epsilon$, for all $x, y \in X$, then

$$\left| \int_X \int_X \overline{g(x)}[K(x, y) - K_N(x, y)]g(y)d\mu(x)d\mu(y) \right| \le$$

$$\epsilon \int_X \int_X \left| \overline{g(x)}g(y) \right| d\mu(x)d\mu(y) \le \epsilon \|g\|_2^2 \mu(X).$$

From this inequality it follows that

$$\langle T_K g, g \rangle = \lim_{N \to +\infty} \int_X \int_X \overline{g(x)} K_N(x, y)g(y)d\mu(y)d\mu(x)$$

$$= \lim_{N \to +\infty} \sum_{n=1}^{N} \int_X \int_X \overline{g(x)}e_n(x)\overline{e_n(y)}g(y)d\mu(y)d\mu(x)$$

$$= \lim_{N \to +\infty} \sum_{n=1}^{N} |\langle g, e_n \rangle|^2 \ge 0.$$

It follows that T_K is a positive operator. $\qquad\qquad\square$

We now prove a partial converse.

Proposition 11.11. *Let $X \subseteq \mathbb{R}^d$ be a compact set, let $K : X \times X \to \mathbb{C}$ be continuous and let μ be a finite Borel measure with support X. If T_K is positive, then K is a positive semidefinite function on X, that is, K is a Mercer kernel.*

Proof. Given distinct points, $x_1, \ldots, x_n$ in X and scalars $\lambda_1, \ldots, \lambda_n$ we must prove that $\sum_{i,j=1}^{n} \overline{\lambda_i}\lambda_j K(x_i, x_j) \ge 0$.

Fix $\epsilon > 0$ and choose disjoint open neighborhoods U_i of x_i such that for $(x, y) \in U_i \times U_j$ we have that $|K(x, y) - K(x_i, x_j)| < \epsilon$. Since the support of μ is X we have that $\mu(U_i) = r_i \ne 0$. One has that

$$\int_{U_i} \int_{U_j} |K(x, y) - K(x_i, x_j)|d\mu(x)d\mu(y) < \epsilon r_i r_j.$$

Let $g = \sum_{j=1}^{n} \lambda_j r_j^{-1} \chi_{U_j}$, then

$$\left| \langle T_K g, g \rangle - \sum_{i,j=1}^{n} \overline{\lambda_i}\lambda_j K(x_i, x_j) \right| \le$$

$$\sum_{i,j=1}^{n} | \int_{U_i} \int_{U_j} \overline{\lambda_i} \lambda_j r_i^{-1} r_j^{-1} K(x,y) d\mu(x) d\mu(y) - \overline{\lambda_i} \lambda_j K(x_i, x_j)| \le n^2 \epsilon.$$

Since $\langle T_K g, g \rangle \ge 0$ and $\epsilon > 0$ was arbitrary, the result follows.   □

Recall that a collection $\mathcal{F}$ of functions on $X \subseteq \mathbb{R}^d$ is called
EQUICONTINUOUS provided that for every $\epsilon > 0$ there is a $\delta > 0$ such
that when $f \in \mathcal{F}$ and $d_2(x, z) < \delta$ then $|f(x) - f(z)| < \epsilon$, where d_2
denotes the usual Euclidean metric on X. The family of functions is called
POINTWISE BOUNDED provided that, for each $x \in X$, we have $\sup\{|f(x)| :
f \in \mathcal{F}\} < +\infty$.

Proposition 11.12. *Let K be a Mercer kernel on X and let μ be a finite Borel
measure. Then $\mathcal{F} = \{T_K g : g \in L^2(X, \mu), \|g\|_2 \le 1\}$ is an equicontinuous,
pointwise bounded family of functions on X.*

Proof. Since K is uniformly continuous on $X \times X$, there exists a $\delta > 0$, such
that if $d_2(x, z) < \delta$ then $|K(x, y) - K(z, y)| < \frac{\epsilon}{\sqrt{\mu(X)+1}}$.
    Hence,

$$|T_K g(x) - T_K g(z)| \le \int_X |(K(x, y) - K(z, y)) g(y)| d\mu(y) \le$$
$$\left(\int_X |K(x, y) - K(z, y)|^2 d\mu(y) \right)^{1/2} \|g\|_2 < \epsilon.$$

Also,

$$|(T_K g)(x)| \le \left(\int_X |K(x, y)|^2 d\mu(y) \right)^{1/2} \|g\|_2$$

and so $\mathcal{F}$ is pointwise bounded.   □

The next two results that we shall need are general facts about operators on
a Hilbert space.

Lemma 11.13. *Let $\mathcal{H}$ be a Hilbert space and let $A : \mathcal{H} \to \mathcal{H}$ be a bounded
operator. If $h \in \mathcal{H}$ is a unit vector with $\|Ah\| = \|A\|$ and $k \perp h$ is any vector,
then $Ak \perp Ah$.*

Proof. We may assume that k is a unit vector and, after multiplying by a
scalar, that $\langle Ah, Ak \rangle \ge 0$. Since $k \perp h$, $\cos(t)h + \sin(t)k$ is a unit vector. Set

$$f(t) = \|A(\cos(t)h + \sin(t)k)\|^2$$

so that $f(t) \le \|A\|^2$ with $f(0) = \|A\|^2$.

Thus, $t = 0$ is a maximum of f and so $0 = f'(0) = 2 \langle Ah, Ak \rangle$ and the result follows. □

Lemma 11.14. *Let $\mathcal{H}$ be a Hilbert space and let $P : \mathcal{H} \to \mathcal{H}$ be a positive operator. If $h \in \mathcal{H}$ is a unit vector such that $\|Ph\| = \|P\|$, then h is an eigenvector with eigenvalue $\|P\|$.*

Proof. Write $Ph = \lambda h + k$ with $k \perp h$. Then $\langle Ph, h \rangle = \lambda$ and so $\lambda \geq 0$. By the above lemma,

$$0 = \langle Pk, Ph \rangle = \langle Pk, \lambda h + k \rangle$$

so that

$$0 \leq \langle Pk, k \rangle = -\lambda \langle Pk, h \rangle = -\lambda \langle k, \lambda h + k \rangle = -\lambda \|k\|^2.$$

Thus, either $\lambda = 0$ or $k = 0$. If $\lambda = 0$, then $P^{1/2}h = 0$ and $Ph = 0$ so that $\|P\| = 0$ and the result follows. If $\lambda \neq 0$, then $k = 0$ and the result follows. □

We can now prove the main theorem of this section.

Theorem 11.15 (MERCER'S THEOREM). *Let K be a Mercer kernel on X, let μ be a finite Borel measure with support X, and let $T_K : L^2(X, \mu) \to L^2(X, \mu)$ be the associated integral operator. Then there is a countable collection of orthonormal continuous functions $\{e_n\}$ on X that are eigenvectors for T_K with corresponding eigenvalues $\{\lambda_n\}$ such that for every $g \in L^2(X, \mu)$ we have*

$$T_K g = \sum_n \lambda_n \langle g, e_n \rangle e_n.$$

Furthermore, $K(x, y) = \sum_n \lambda_n e_n(x) \overline{e_n(y)}$.

Proof. We assume that $K \neq 0$ or there is nothing to prove.

Choose a sequence of unit vectors $\{g_n\}$ in $L^2(X, \mu)$ with $\lim_n \|T_K^2 g_n\| = \|T_K^2\|$. Note that since T_K is positive, $\|T_K^2\| = \|T_K\|^2$. By Proposition 11.12 the sequence of continuous functions $T_K g_n$ is pointwise bounded and equicontinuous. Hence, by the Arzelà–Ascoli Theorem, there is a subsequence that converges uniformly to a continuous function f. Relabeling the original sequence, we may assume that $T_K g_n$ converges uniformly to f on X. Since μ is a finite Borel measure, uniform convergence implies that $\|T_K g_n - f\|_2 \to 0$ and $\|f\|_2 \leq \|T_K\|$. Hence,

$$\|T_K\| \|f\|_2 \geq \|T_K f\|_2 = \lim_n \|T_K (T_K g_n)\|_2 = \|T_K\|^2 \geq \|T_K\| \|f\|_2$$

and so we have equality throughout. Thus, $\|T_K\|^2 = \|T_K f\|_2 = \|T_K\|\|f\|_2$ and $\|f\|_2 = \|T_K\|$.

Set $e_1 = \frac{f}{\|f\|_2}$, then $\|e_1\|_2 = 1$ and $\|T_K e_1\| = \|T_K\|$. Hence, by the above lemma, e_1 is an eigenvector with eigenvalue $\lambda_1 = \|T_K\|$. If we let P_1 be the orthogonal projection onto the span of e_1, so that $P_1 g = \langle g, e_1 \rangle e_1$, then $R_1 = T_K - \lambda_1 P_1$ is a positive operator and $\|R_1\| \leq \|T_K\|$. Moreover,

$$R_1 g = \int_X K(x, y)g(y) - \lambda_1 e_1(x)\overline{e_1(y)}g(y)d\mu(y),$$

so we see that R_1 is an integral operator with symbol

$$K_1(x, y) = K(x, y) - \lambda_1 e_1(x)\overline{e_1(y)}.$$

Note that K_1 is continuous and

$$T_{K_1} e_1 = T_K e_1 - \lambda_1 P_1 e_1 = 0.$$

Since T_{K_1} is a positive operator, K_1 is a positive semidefinite function on X by Proposition 11.11.

Thus K_1 is a Mercer kernel. Also, $\|T_K\| \geq \|T_{K_1}\|$. If $\|T_{K_1}\| = 0$, then $K(x, y) = \lambda_1 e_1(x)\overline{e_1(y)}$ and we are done. Otherwise, we may repeat the argument to obtain a continuous function e_2 with $\|e_2\|_2 = 1$ and $T_{K_1} e_2 = \lambda_2 e_2$ with $\lambda_2 = \|T_{K_1}\|$. Since

$$\lambda_2 \langle e_2, e_1 \rangle = \langle T_{K_1} e_2, e_1 \rangle = \langle e_2, T_{K_1} e_1 \rangle = 0,$$

we have $e_1 \perp e_2$. Hence, $T_K e_2 = T_{K_1} e_2 + \lambda_1 P_1 e_2 = \lambda_2 e_2$.

Proceeding in this manner, we produce unit eigenvectors $\{e_n\}$ and a decreasing sequence of eigenvalues $\{\lambda_n\}$ for T_K such that for each N,

$$K_N(x, y) = K(x, y) - \sum_{n=1}^{N} \lambda_n e_n(x)\overline{e_n(y)},$$

is a Mercer kernel and $\lambda_{N+1} = \|T_{K_N}\|$. This process either terminates after finitely many steps, in which case $K(x, y) = \sum_{n=1}^{N} \lambda_n e_n(x)\overline{e_n(y)}$ or we produce a sequence of eigenvectors and eigenvalues in this fashion.

In the case that we produce a sequence, by Proposition 11.12 some subsequence of the vectors $T_K e_{n_j} = \lambda_{n_j} e_{n_j}$ converges uniformly and hence in norm. Thus, the vectors in the subsequence are Cauchy. But since these vectors are orthonormal, we have that $\|\lambda_{n_j} e_{n_j} - \lambda_{n_k} e_{n_k}\|_2 = \sqrt{|\lambda_{n_j}|^2 + |\lambda_{n_k}|^2}$ and hence $\lambda_{n_j} \to 0$ for the subsequence. Since the sequence is decreasing, we have $\|T_{K_n}\| = \lambda_{n+1} \to 0$.

This last fact yields that

$$K(x, y) = \sum_{n=1}^{+\infty} \lambda_n e_n(x) \overline{e_n(y)}.$$

$\square$

When the support of μ is smaller than X, then one gets, essentially, the same result. To see this, let $S \subseteq X$ be the support of μ, then S is a compact set and the restriction of K to S is a Mercer kernel on S and any integral has the same value if it is only taken over S. Thus, we get continuous eigenfunctions on S and all of the above conclusions on S. If we extend these eigenfunctions to X by setting them to be 0 on $X \backslash S$, then these are still eigenfunctions for the integral operator defined on X. Since $\mu(X \backslash S) = 0$, we conclude that the eigenfunctions are continuous almost everywhere and that they sum to K almost everywhere.

Let K be a Mercer kernel on X and let T_K be the associated integral operator. We would like to describe the relationship between the RKHS $\mathcal{H}(K)$ and the RKHS that is the range of T_K.

From our earlier results, we already know that $K^{(2)} = K \square K$ is the kernel function for the RKHS that is the range space of T_K.

Proposition 11.16. *Let K be a Mercer kernel on X and let μ be a finite Borel measure on X, then T_K^2 is an integral operator with symbol $K^{(2)}(x, y) = \int_X K(x, t) K(t, y) \, d\mu(t)$.*

Proof. To prove this result we calculate the action of T_K^2:

$$T_K^2 f(x) = \int_X K(x, t) T_K f(t) \, d\mu(t)$$

$$= \int_X K(x, t) \int_X K(t, y) f(y) \, d\mu(y) \, d\mu(t)$$

$$= \int_X f(y) \left(\int_X K(x, t) K(t, y) \, d\mu(t) \right) d\mu(y).$$

$\square$

Proposition 11.17. *Let K be a Mercer kernel on X, let μ be a finite Borel measure on X, and set $C = \int_X K(t, t) d\mu(t)$. Then $CK - K^{(2)}$ is a Mercer kernel on X and the range of T_K is contained in $\mathcal{H}(K)$.*

Proof. We already know that $K^{(2)}$ is continuous. We only need to prove that $CK - K^{(2)}$ is a kernel function on X. Note that by Cholesky,

$$L_t(x, y) = K(t, t)K(x, y) - K(x, t)K(t, y)$$

$$= K(t, t)\left(K(x, y) - \frac{K(x, t)K(t, y)}{K(t, t)}\right)$$

is a kernel function, for each fixed value of t. In particular, this expression is the kernel function for the set of functions in $\mathcal{H}(K)$ that vanish at t. Assume that $\mu(X) = 1$.

To see that $CK \geq K^{(2)}$. We have

$$K(x, y)\int_X K(t, t)\,d\mu(t) - \int_X K(x, t)K(t, y)d\mu(t) = \int_X L_t(x, y)\,d\mu(t).$$

Given points $x_1, \ldots, x_n \in X$ and scalars $\alpha_1, \ldots, \alpha_n$ we have

$$\sum_{i,j=1}^n \alpha_i\overline{\alpha_j}(CK(x_i, x_j) - K^{(2)}(x_i, x_j)) = \int_X \sum_{i,j=1}^n \alpha_i\overline{\alpha_j}L_t(x_i, x_j)\,d\mu(t).$$

Since L_t is a kernel for each t the expression in the integrand is nonnegative for each t and hence the integral is nonnegative.

Finally, since $K^{(2)} \leq CK$, we have that $\mathcal{R}(T_K) = \mathcal{H}(K^{(2)}) \subseteq \mathcal{H}(K)$, by Theorem 5.1.   □

What is remarkable about the above result is that $\mathcal{R}(T_K)$ depends upon the measure μ while $\mathcal{H}(K)$ is independent of μ!

We have seen that the range space of the integral operator T_K is an RKHS with kernel $K^{(2)}$. We also have that T_K^2 is an integral operator with kernel $K^{(2)}$. Recall that given a positive operator T, there exists a unique positive square root $T^{1/2}$. Thus, if T_K is an integral operator, then $T_K^{1/2}$ exists, but may not be an integral operator. Motivated by this observation, and earlier result, we now show that $\mathcal{H}(K)$ is the range space of the operator $T_K^{1/2}$. Again, this is surprising, since $\mathcal{R}(T_K^{1/2})$ is defined using the measure, while $\mathcal{H}(K)$ is independent of any measure.

Theorem 11.18. *Let K be a Mercer kernel on X, let μ be a finite Borel measure with support X and let $\{e_n\}$ be the orthonormal family of eigenvectors corresponding to non-zero eigenvalues $\{\lambda_n\}$ of T_K given by Mercer's Theorem. Then:*

1. *$S(x, y) = \sum_n \sqrt{\lambda_n}e_n(x)e_n(y)$ is the symbol of a bounded integral operator on $L^2(X, \mu)$;*
2. *T_S is a positive operator with $T_S^2 = T_K$;*
3. *$\mathcal{R}(T_S) = \mathcal{H}(K)$;*
4. *T_S defines an isometry of $\mathcal{K}(T_K)^\perp$ onto $\mathcal{H}(K)$;*

5. $\{\sqrt{\lambda_n}e_n\}$ *is an orthonormal basis for* $\mathcal{H}(K)$;
6. $\mathcal{H}(K)$ *consists precisely of the functions* $\sum_n \alpha_n e_n(x)$ *such* $\sum_n \frac{|\alpha_n|^2}{\lambda_n} < \infty$.

Proof. Because the functions $e_n(y)$ are orthogonal in $L^2(X, \mu)$, for each $x \in X$, $\|S(x, \cdot)\|_2 = \sum_n \lambda_n |e_n(x)|^2 = K(x, x)$ by Mercer's theorem. Hence, for each x, $S(x, \cdot) \in L^2(X, \mu)$ and so there is a well-defined integral operator T_S from $L^2(X, \mu)$ into a space of functions on X. For any $g \in L^2(X, \mu)$, $T_S g(x) = \sum_n \sqrt{\lambda_n} \langle g, e_n \rangle e_n(x)$. Thus, $\|T_S g\|_2^2 = \sum_n \lambda_n |\langle g_n, e_n \rangle|^2 \leq C \|g\|_2$, where $C = \sup_n \lambda_n < \infty$. Hence, T_S defines a bounded integral operator on $L^2(X, \mu)$.

A similar calculation shows that $\langle T_S g, g \rangle = \sum_n \sqrt{\lambda_n} |\langle g, e_n \rangle|^2 \geq 0$, so T_S is a positive operator. Also that $T_S^2 g = \sum_n \lambda_n \langle g, e_n \rangle e_n = T_K g$. Hence, T_S is the positive square root of T_K.

By our earlier results, we know that the range of T_S is the reproducing kernel Hilbert space with kernel,

$$S \square S^*(x, z) = \int_X S(x, y)\overline{S(y, z)}d\mu(y) =$$

$$\int_X \sum_{n,m} \sqrt{\lambda_n}\sqrt{\lambda_m}e_n(x)\overline{e_n(y)}e_m(y)\overline{e_m(z)}d\mu(y) =$$

$$\sum_n \lambda_n e_n(x)\overline{e_n(z)} = K(x, z).$$

We have shown earlier that every integral operator T_S defines an isometry from $\mathcal{K}(T_S)^{\perp}$ onto $\mathcal{H}(S \square S^*)$. Now check that $\mathcal{K}(T_S)^{\perp} = \text{span}\{e_n\}^- = \mathcal{K}(T_K)^{\perp}$. Thus, T_S defines a unitary operator from $span\{e_n\}^-$ onto $\mathcal{H}(K)$. Since $\{e_n\}$ are an orthonormal basis for $\mathcal{K}(T_K)^{\perp}$ their images, $\{\sqrt{\lambda_n}e_n\}$ are an orthonormal basis for $\mathcal{H}(K)$.

The last conclusion follows easily from the fact that T_S is an onto isometry.

□

11.3.1 An application of Mercer's theory

Let us consider the case of the kernel $K(x, y) = \min\{x, y\}$. This defines a Mercer kernel on any compact subset $X \subseteq [0, +\infty)$. Let's consider the case $X = [0, b]$ for some $b > 0$.

When we consider the whole half-line

$$\mathcal{H}(K) = \{f : f(0) = 0, f \text{ absolutely continuous}, f' \in L^2\}.$$

Thus, by our restriction Theorem 5.8 the RKHS generated on $[0, b]$ by K will just be the restriction of such functions to $[0, b]$.

Now if we take *any* finite Borel measure μ with support $[0, b]$ and consider the integral operator

$$T_K g(x) = \int_{[0,b]} \min\{x, y\} g(y) d\mu(y),$$

then this will be a positive operator with a basis of eigenfunctions. Moreover, when we wish to seek eigenfunctions $\{e_n\}$ for this operator, we know *a priori* that every eigenfunction will be an absolutely continuous function satisfying $f(0) = 0$ and whose derivative will be square-integrable with respect to Lebesgue measure on $[0, b]$. Finally, we also know that these eigenfunctions will be orthogonal in $L^2(X, \mu)$ and in $\mathcal{H}(K)$. This means that the eigenfunctions will have to satisfy,

$$\int_{[0,b]} e_n(y)\overline{e_m(y)} d\mu(y) = 0 \text{ and } \int_0^b e_n'(y)\overline{e_m'(y)} dy = 0,$$

for $n \neq m$.

11.4 Square integrable symbols

This section requires some familiarity with measure theory and functional analysis.

A widely studied family of integral operators are those whose symbol is square integrable with respect to product measure. In this section we show that these correspond to the Hilbert–Schmidt operators. We shall assume that μ is a σ-finite measure on X so that there is a product measure $\mu \times \mu$ on $X \times X$. If $S : X \times X \to \mathbb{C}$ is square integrable with respect to $\mu \times \mu$ then

$$\int_{\mu \times \mu} |S(x, y)|^2 d(\mu \times \mu) = \int_X \left(\int_X |S(x, y)|^2 d\mu(y) \right) d\mu(x) =$$

$$\int_X \left(\int_X |S(x, y)|^2 d\mu(x) \right) d\mu(y)$$

and hence

$$\int_X |S(x, y)|^2 d\mu(x) < +\infty \text{ a.e. } y \text{ and } \int_X |S(x, y)|^2 d\mu(y) < +\infty \text{ a.e. } x.$$

It is a standard exercise in measure theory to show that there exists a measurable subset $Y \subseteq X$ such that $\mu(X \backslash Y) = 0$ and these two integrals are finite for all $x, y \in Y$. Moreover, by setting

$$\tilde{S}(x, y) = \begin{cases} S(x, y), & (x, y) \in Y \times Y \\ 0, & (x, y) \notin Y \times Y \end{cases}$$

one obtains a function that is equal to S almost everywhere with respect to $\mu \times \mu$ and has the property that each partial integral is bounded everywhere (Exercise (11.2).

Since in the theory of RKHSs we are interested in functions that exist everywhere, we shall assume that $S : X \times X \to \mathbb{C}$ is square integrable with respect to the product measure and has the property that

$$\int_X |S(x, y)|^2 d\mu(y) < +\infty \, \forall x \text{ and } \int_X |S(x, y)|^2 d\mu(x) < +\infty \, \forall y.$$

We shall call such a function a standard square integrable function on the product domain.

When S is a standard square integrable function on the product domain, then for each $x \in X$, we have that $h_x(\cdot) = S(x, \cdot) \in L^2(X)$ and hence for every $x \in X$ and $g \in L^2(X)$ the value $(T_h g)(x) = \int_X S(x, y)g(y)d\mu(y)$ is defined.

Moreover, since

$$|(T_S g)(x)|^2 \le \left(\int_X |S(x, y)|^2 d\mu(y)\right)\left(\int_X |g(y)|^2 d\mu(y)\right),$$

we have that

$$\|T_S g\|_2^2 \le \|g\|_2^2 \left(\int_X \int_X |S(x, y)|^2 d\mu(y) d\mu(x)\right) = \|g\|_2^2 \|S\|_2^2.$$

Thus, the integral operator T_S maps $L^2(X)$ into $L^2(X)$ and is bounded with $\|T_S\| \le \|S\|_2$.

So beginning with a standard square integrable function S on the product domain, we obtain a bounded integrable operator, $T_S : L^2(X) \to L^2(X)$. Perhaps more importantly, the range of this integral operator is not just a subspace of $L^2(X)$ but it is a vector space of actual functions on X. We have seen, more generally, that there is a natural norm on this space that makes it into an RKHS of functions on X and calculated the reproducing kernel in terms of the function S. Namely,

$$K(x, z) = S \square S^*(x, z) = \int_X S(x, y)\overline{S(y, z)} d\mu(y).$$

We leave the proof of the next fact to the exercises.

Proposition 11.19. *Let (X, μ) be a σ-finite measure space and let S_1 and S_2 be standard square integrable functions on the product domain. Then:*

- *$S = S_1 \square S_2$ is a standard square integrable function on the product domain;*
- *$T_S = T_{S_1} \circ T_{S_2}$;*
- *$(T_{S_1})^* = T_{S_1^*}$.*

An operator $T \in B(H)$ is called a $\boxed{\text{HILBERT–SCHMIDT OPERATOR}}$, if for any orthonormal basis e_n the sum $\sum_{n=1}^{\infty} \|Te_n\|^2$ is finite. It is a standard result in operator theory that a Hilbert–Schmidt operator is compact, and that the above sum is independent of the choice of orthonormal basis. For a Hilbert–Schmidt operator T, the Hilbert–Schmidt norm of the operator is defined to be $\sum_{n=1}^{\infty} \|Te_n\|^2$.

Proposition 11.20. *Let μ be a sigma-finite measure on X and let S be a standard square-integrable function on $X \times X$. Then the associated integral operator is Hilbert–Schmidt and $\|T_S\|_{HS}^2 = \iint_{X \times X} |S(x, y)|^2 \, dx \, dy$.*

Proof. Let e_n be an orthonormal basis of $L^2(X)$. We have,

$$\int_X |S(x, y)|^2 \, d\mu(y) = \sum_{n=1}^{\infty} |\langle S(x, \cdot), e_n \rangle|^2$$

$$= \sum_{n=1}^{\infty} \left| \int_X S(x, t) e_n(t) \, d\mu(t) \right|^2$$

$$= \sum_{n=1}^{\infty} |T_S e_n(x)|^2.$$

We now integrate with respect to x to get

$$\int_X \int_X |K(x, y)|^2 \, d\mu(y) \, d\mu(x) = \sum_{n=1}^{\infty} \|T_S e_n\|^2.$$

This establishes our claim. $\qquad \square$

Now, conversely, assume that $T : L^2(X, \mu) \to L^2(X, \mu)$ is a Hilbert–Schmidt operator and that $L^2(X, \mu)$ is a separable Hilbert space. Fix an orthonormal basis $\{e_n(y)\}$ for $L^2(X, \mu)$ of square-integrable functions on X and or each n, choose a square-integrable function $\{f_n\}$ so that $Te_n = f_n$. Then for any $g = \sum_n \alpha_n e_n$ we have that $Tg = \sum_n \alpha_n f_n$ and since $\sum_n \|f_n\|_2^2 < \infty$, this latter series converges in norm. Set

$$S(x, y) = \sum_n f_n(x) \overline{e_n(y)}.$$

We leave it to the reader to check that this function is square-integrable on $X \times X$. Thus, we may modify all these functions on a set of measure 0 so that it is a standard square-integrable function.

Proposition 11.21. *Let S be defined as above. Then $T = T_S$.*

We leave the proof to the reader.

The above two results show that the Hilbert–Schmidt operators on a separable L^2 space are precisely the integral operators with square-integrable symbol. Since every separable Hilbert space is unitarily equivalent to a separable L^2 spaces and the property of being a Hilbert–Schmidt operator is preserved under unitary equivalence, we see that the study of Hilbert–Schmidt operators is essentially the study of integral operators with square-integrable symbol.

11.5 Exercises

Exercise 11.1. Let $\mathcal{H}$ and $\mathcal{K}$ be Hilbert spaces, let $T : \mathcal{H} \to \mathcal{K}$ be a bounded operator, and let $\mathcal{R}(T)$ be equipped with the range norm.

- Prove that $T|_{\mathcal{K}(T)^\perp} : \mathcal{K}(T)^\perp \to \mathcal{R}(T)$ is one-to-one, onto and an isometry.
- Let $S = \left(T|_{\mathcal{K}(T)^\perp}\right)^{-1} : \mathcal{R}(T) \to \mathcal{K}(T)^\perp$ and for $x, y \in \mathcal{R}(T)$ define $\langle x, y \rangle_{\mathcal{R}(T)} = \langle S(x), S(y) \rangle_{\mathcal{H}}$. Prove that this is an inner product on $\mathcal{R}(T)$ and that $\mathcal{R}(T)$ is a Hilbert space in this inner product.

Exercise 11.2. Let (X, μ) be a σ-finite measure space and let $S : X \times X \to \mathbb{C}$ be square integrable with respect to the product measure $\mu \times \mu$. Prove that there exists a measurable subset $Y \subseteq X$ with $\mu(X \backslash Y) = 0$ such that $\int_X |S(x, y)|^2 d\mu(y) < +\infty$ for all $x \in Y$ and $\int_X |S(x, y)|^2 d\mu(x) < +\infty$ for all $y \in Y$. Prove that the function $\tilde{S}$ that is defined to be equal to S on $Y \times Y$ and 0 on $(X \times X) \backslash (Y \times Y)$ is equal to S almost everywhere with respect to $\mu \times \mu$ and has the property that both of these partial integrals are finite everywhere, i.e. that $\tilde{S}$ is a standard square integrable function.

Exercise 11.3. Prove Proposition 11.19.

Exercise 11.4. Prove Proposition 11.21 and the preceding claims about S.

Exercise 11.5. Let m be Lebesgue measure on $[0, 1]$ and let $T_K g(x) = \int_0^1 K(x, y) g(y) dy$, where $K(x, y) = min\{x, y\}$. Apply Mercer's theory to find the eigenfunctions and eigenvalues of this operator.

Exercise 11.6. Let μ be the measure on $[0, 1]$ that is Lebesgue measure plus the measure that is the point mass at $1/2$ so that $\int_0^1 g(y) d\mu(y) = \int_0^1 g(y) dy + g(1/2)$. For $K(x, y) = min\{x, y\}$, compute the eigenfunctions and eigenvalues.

Exercise 11.7. Let K be a kernel on the set T and suppose that there is a measure space (X, m), and a function $\phi : T \times X \to \mathbb{R}$ such that for each $t \in T$, $\phi_t(\cdot) = \phi(t, \cdot \in L^2(X, m)$. If $K(s, t) = \int_X \phi_s(x)\phi_t(x)\, dm(x)$, then prove that $\mathcal{H}(K)$ is isomorphic to the space of functions of the form $f(t) = \int_X \phi(t, x)g(x)\, dm(x)$, where g is in the closed span of functions of the form $\{\phi(t, \cdot) : t \in T\} \subseteq L^2(X, m)$.

Exercise 11.8. Let $\phi_1, \ldots, \phi_n$ be continuous functions on a compact set $X \subseteq \mathbb{R}^d$ and let μ be a Borel measure on X. Let $K(x, y) = \sum_{i=1}^n \phi_i(x)\overline{\phi_i(y)}$. Show that the integral operator T_K has rank at most n.

12

Stochastic processes

In this chapter we look at the role of RKHS in probability and stochastic processes. These results are widely used in many areas, including machine learning and the statistical analysis of time series.

We will briefly review the terminology and notation from probability and measure theory that we need. However, we assume that the reader is familiar with these areas and cite many facts without proofs. Much of this material can be found in [11].

12.1 Covariance

The central idea that we will be concerned with is the notion of covariance. This is the primary link between RKHS and stochastic processes.

Let (Ω, P) be a probability space. A $\boxed{\text{RANDOM VARIABLE}}$ is a measurable function $X : \Omega \to \mathbb{R}$. We assume throughout this chapter that all random variables are square-integrable, and consequently, integrable. Given a random variable $X : \Omega \to \mathbb{R}$, the $\boxed{\text{EXPECTED VALUE}}$, or $\boxed{\text{MEAN}}$, of X is given by

$$\mathbb{E}[X] = \int_\Omega X \, dP.$$

The $\boxed{\text{VARIANCE}}$ of a random variable X is given by

$$\text{Var}[X] = \mathbb{E}[(X - \mathbb{E}[X])^2] = \mathbb{E}[X^2] - \mathbb{E}[X]^2.$$

The variance is a measure of dispersion of the random variable from its expected value. On the other hand the covariance of two random variables X, Y is a measure of their similarity. The $\boxed{\text{COVARIANCE}}$ between X, Y is given by

$$\text{Cov}(X, Y) = \mathbb{E}[(X - \mathbb{E}[X])(Y - \mathbb{E}[Y])] = \mathbb{E}[XY] - \mathbb{E}[X]\mathbb{E}[Y].$$

When $\text{Cov}(X, Y) = 0$ we say that X and Y are $\boxed{\text{UNCORRELATED}}$. Writing the expression for the covariance as an integral we see that

$$\text{Cov}(X, Y) = \int_{\Omega} (X - \mathbb{E}[X])(Y - \mathbb{E}[Y]) \, dP.$$

Note that $\text{Cov}(X, X) = \text{Var}[X]$. One can check that $\text{Cov}(X, Y) = \text{Cov}(X - \mathbb{E}[X], Y - \mathbb{E}[Y])$. Since $X - \mathbb{E}[X]$ and $Y - \mathbb{E}[Y]$ are random variables with expected values 0, we often reduce the clutter in arguments by assuming that the expected value of our random variables is 0. When both random variables have 0 expected value, then the covariance is simply $\text{Cov}(X, Y) = \mathbb{E}[XY]$.

Given random variables $X_1, \ldots, X_n$ we can form the $n \times n$ covariance matrix $(\text{Cov}(X_i, X_j))$. Our next result shows that this matrix is positive. This is the first step in bridging the areas of stochastic processes and RKHS.

Proposition 12.1. *Let $X_1, \ldots, X_n$ be random variables, then the matrix $C = (\text{Cov}(X_i, X_j))$ is positive.*

Proof. We can assume that $\mathbb{E}[X_i] = 0$ for $i = 1, \ldots, n$. In this case we see that $\text{Cov}(X_i, X_j) = \langle X_i, X_j \rangle_{L^2(\Omega)}$. Since the Grammian matrix $(\langle x_i, x_j \rangle_{\mathcal{H}})$ of any set of vectors $x_1, \ldots, x_n$ in a Hilbert space is positive, we see that $C \geq 0$. $\square$

Often a set of random variables, $X_1, \ldots, X_n$ on Ω is thought of as a $\boxed{\text{VECTOR-VALUED RANDOM VARIABLE}}$, $X : \Omega \to \mathbb{R}^n$. In this case the expected value is the vector, $\mathbb{E}[X] = (\mathbb{E}[X_1], \ldots, \mathbb{E}[X_n])$ and the matrix C is called the covariance of X. More formally, given random variables $X_1, \ldots, X_n, Y_1, \ldots, Y_n$, we define $X := (X_1, \ldots, X_n)^T$ and $Y := (Y_1, \ldots, Y_n)^T$ to be column vectors. Then $XY^T = (X_i Y_j)$ is a matrix of random variables and one defines $\text{Cov}(X, Y) := (\text{Cov}(X_i, Y_j)) = \mathbb{E}[XY^T] - \mathbb{E}[X]\mathbb{E}[Y]^T$.

Given a random variable $X : \Omega \to \mathbb{R}$ we define the $\boxed{\text{DISTRIBUTION}}$ $\boxed{\text{FUNCTION}}$ $F_X : \mathbb{R} \to [0, 1]$ by $F_X(a) = P(X \leq a)$. It is well-known that F_X is a right-continuous, non-decreasing function. Conversely, every right-continuous, non-decreasing function F with $\lim_{x \to -\infty} F(x) = 0$ and $\lim_{x \to +\infty} F(x) = 1$ is the distribution function of a random variable.

Two random variables $X, Y : \Omega \to \mathbb{R}$ are called $\boxed{\text{INDEPENDENT}}$ if, for all scalars $a, b \in \mathbb{R}$, we have

$$P(X \leq a, Y \leq b) = P(X \leq a)P(Y \leq b).$$

Note that two independent random variables are uncorrelated, but that the converse is false, in general.

A collection $\{X_i : i \in I\}$ of random variables on Ω is called $\boxed{\text{INDEPENDENT}}$ $\boxed{\text{AND IDENTICALLY DISTRIBUTED}}$ (usually abbreviated i.i.d.) if $F_{X_n} = F_{X_m}$ and X_n, X_m are independent for all pairs $m, n \in I$ such that $m \neq n$.

Given a distribution F and an index set I, it is known that there always exists a collection of i.i.d. random variables $\{X_i : i \in I\}$ each with distribution F. It is the existence of such i.i.d. random variables, and not their construction, that is important for our exposition.

12.2 Stochastic processes

We have seen that the covariance matrix of a vector-valued random variable is positive. In this section we introduce stochastic processes, which generalize the notion of vector-valued random variables, and we show that the covariance function of such a process is a kernel function.

To see how to do this generalization, note that a vector-valued random variable $(X_1, \ldots, X_n) : \Omega \to \mathbb{R}^n$ can be viewed as a function $X : T \times \Omega \to \mathbb{R}$ where $T = \{1, \ldots, n\}$ by setting $X(i, \omega) = X_i(\omega)$. In a stochastic process the set T is allowed to be more general.

Definition 12.2. Let T be a set and let (Ω, P) be a probability space. A $\boxed{\text{STOCHASTIC PROCESS}}$ is a function $X : T \times \Omega \to \mathbb{R}$ such that for each fixed $t \in T$ the function $X_t : \Omega \to \mathbb{R}$ given by $X_t(\omega) = X(t, \omega)$ is a random variable.

For each fixed $\omega \in \Omega$ the function $t \to X(t, \omega)$ is called a $\boxed{\text{TRAJECTORY}}$ or a $\boxed{\text{SAMPLE PATH}}$ of the process. Usually, this function is denoted $X(\cdot, \omega)$.

In the case that T is a metric space or a measure space, one often wants to assume that the sample paths are measurable or continuous. We will touch upon these issues later.

Recall that throughout this chapter we are assuming that all random variables are square-integrable.

It is often the case that T represents a time parameter, in which case T is usually taken to be an interval, or $\mathbb{N}$ in the case of discrete time.

Associated to a stochastic process X are two functions: the mean $m : T \to \mathbb{R}$ and the covariance $K : T \times T \to \mathbb{R}$.

Definition 12.3. The $\boxed{\text{MEAN FUNCTION}}$ of a stochastic process X is given by $m_X(t) = \mathbb{E}[X_t]$. The $\boxed{\text{COVARIANCE FUNCTION}}$ of a stochastic process is given by $K_X(s, t) = \text{Cov}(X_s, X_t)$.

When there is no ambiguity we will suppress the subscript in the notation for the mean and covariance.

The next result shows that we can always replace X by a stochastic process with zero mean without changing its covariance.

Proposition 12.4. *Let X be a stochastic process with mean function m_X. Then the process $Y_t(\omega) = X_t(\omega) - m_X(t)$ has the following properties:*

1. $m_Y(t) = 0$ *for all $t \in T$;*
2. $\text{Cov}(Y_s, Y_t) = \text{Cov}(X_s, X_t)$.

We leave the proof as an exercise.

From our earlier observation Proposition 12.1 we see that for any finite set of points $t_1, \ldots, t_n \in T$, the matrix $(K(t_i, t_j)) \geq 0$. Hence, the covariance function of a stochastic process is a kernel function on T. Therefore, we can associate the RKHS $\mathcal{H}(K)$ to the process X. This insight connects stochastic processes and RKHS theory. We now state this central result.

Theorem 12.5. *Let X be a stochastic process. Then the covariance function of X is a kernel function on T.*

Definition 12.6. Let X be a stochastic process. The RKHS of X is the Hilbert space of functions on T with kernel given by $K(s, t) = \text{Cov}(X_s, X_t)$.

We now highlight some connections between the stochastic process and the associated RKHS. After proving these general statements we will later analyze the $\boxed{\text{WIENER PROCESS}}$, which is a stochastic process with kernel $K(s, t) = \min\{s, t\}$.

To begin, by recalling the results of the last chapter, we can give an explicit description of the RKHS $\mathcal{H}(K)$ of a stochastic process.

Let (Ω, P) be our probability space and $X : T \times \Omega \to \mathbb{R}$ be our stochastic process, and assume that we have normalized by subtracting off the mean so that $m_X(t) = 0$. Since for each fixed $t \in T$, we have that $X(t, \cdot) \in L^2(\Omega, P)$ we have a well-defined integral operator defined for $g \in L^2(\Omega, P)$ which we will denote by

$$I_X(g)(t) = \int_\Omega X(t, \omega) g(\omega) dP(\omega).$$

By Theorem 11.3 the range of this map is exactly the RKHS $\mathcal{H}(K)$. Moreover, if we let $\mathcal{L} = \mathcal{K}(I_X)^\perp$, then we saw in the proof that $I_X : \mathcal{L} \to \mathcal{H}(K)$

acts as an onto isometry, i.e. a unitary. Applying Corollary 11.4 and using the fact that $X(t, \omega)$ is real-valued, we have:

Theorem 12.7. *Let X be a stochastic process on $T \times \Omega$, with mean function $m \equiv 0$. Let K be the covariance function and let $\mathcal{L}$ denote the closed subspace of $L^2(\Omega)$ spanned by $\{X_t : t \in t\}$. Then there is a unitary $U : \mathcal{H}(K) \to \mathcal{L}$ such that $U(k_t) = X_t$.*

We now obtain a representation of the elements in $\mathcal{H}(K)$ as "averages" of the process.

Proposition 12.8. *Let X_t, $\mathcal{L} \subseteq L^2(\Omega)$ be as in 12.7. Then every $f \in \mathcal{H}(K)$ is of the form $f(t) = \mathbb{E}[gX_t]$ and, moreover, there exists a unique function $g \in \mathcal{L}$ such that $f(t) = \mathbb{E}[gX_t]$.*

Proof. Theorem 12.7 tells us that the map $U : \mathcal{H}(K) \to \mathcal{L}$ is a unitary. Let $g = U(f)$. Now,

$$f(t) = \langle f, k_t \rangle = \langle U(f), U(k_t) \rangle = \langle g, X_t \rangle = \int_\Omega gX_t \, dP = \mathbb{E}[gX_t].$$

$\square$

Note that the above result tells us that the map $U^* = U^{-1} : \mathcal{L} \to \mathcal{H}(K)$ is given by $U^*g = \mathbb{E}[gX_t]$.

The results above assume that K is the covariance of a given stochastic process. In the next section we show how to start with a kernel function $K : T \times T \to \mathbb{R}$ and construct a stochastic process with covariance kernel K. Moreover this can be done in such a way that each random variable X_t belongs to a special family known as the Gaussian random variables.

12.3 Gaussian random variables

In this section we begin by reviewing a few facts about random variables that have a Gaussian distribution. Gaussian random variables have many remarkable properties. In particular, the mean and covariance of a Gaussian random variable completely characterize its distribution. Furthermore, unlike general random variables, two Gaussian random variables are independent if and only if they are uncorrelated.

We begin with a brief overview of Gaussian random variables. A random variable Z on a probability space (Ω, P) is called STANDARD GAUSSIAN if its distribution function satisfies

$$F_Z(a) := P(Z \le a) = \frac{1}{\sqrt{2\pi}} \int_{-\infty}^{a} e^{-x^2/2} \, dx.$$

A standard Gaussian has a mean of 0 and a variance of 1. A random variable X has a Gaussian (or Normal) distribution with parameters (μ, σ) if and only if

$$F_X(a) = \frac{1}{\sqrt{2\pi}\sigma} \int_{-\infty}^{a} e^{-\frac{(x-\mu)^2}{2\sigma^2}} \, dx.$$

When this is the case, we will write $X \sim N(\mu, \sigma)$. It is not hard to see that $X \sim N(\mu, \sigma)$ if and only if $\frac{X-\mu}{\sigma}$ is a standard Gaussian. Hence, $\mathbb{E}[X] = \mu$ and $\text{Var}[X] = \sigma^2$.

We extend the notion of a Gaussian random variable to the vector-valued setting.

Definition 12.9. A vector-valued random variable $X = (X_1, \ldots, X_n)^T$ is said to have a $\boxed{\text{JOINT GAUSSIAN DISTRIBUTION}}$ if and only if, for every vector $\alpha = (\alpha_1, \ldots, \alpha_n)^T \in \mathbb{R}^n$, the random variable $\alpha_1 X_1 + \cdots + \alpha_n X_n$ is normally distributed. Given a metric space T and a probability space (Ω, P) a stochastic process $X : T \times \Omega \to \mathbb{R}$ is called a $\boxed{\text{GAUSSIAN PROCESS}}$ if for every finite set of distinct points $\{t_1, \ldots, t_n\} \subseteq T$ the vector-valued random variable $(X_{t_1}, \ldots, X_{t_n})$ has a joint Gaussian distribution.

We let $\mu_i = \mathbb{E}[X_i]$ and let $\mu = (\mu_1, \ldots, \mu_n)^T$ denote the mean of the Gaussian random variable X. We let $\Sigma = (\text{Cov}(X_i, X_j))$ denote the covariance. It is easily shown that $\alpha_1 X_1 + \cdots + \alpha_n X_n$ has a variance of $\alpha^T \Sigma \alpha$.

If the covariance matrix Σ is invertible, then the distribution of X has a density with respect to n-dimensional Lebesgue measure, and we have

$$P(X \in E) = \frac{1}{\sqrt{2\pi \det(\Sigma)}^n} \int_E \exp\left(-(x - \mu)^T \Sigma^{-1}(x - \mu)/2\right) dx,$$

where E is an open set in $\mathbb{R}^n$. An n-dimensional Gaussian is called standard when $\mu = 0$ and $\Sigma = I_n$.

There is a well-known way to construct a vector-valued Gaussian with a prescribed mean and covariance. This can be found in [11] or done as an exercise.

Lemma 12.10. *Let Σ be a positive matrix and suppose that $\Sigma = AA^T$. Let $A = [a_{i,j}]$ and define $X_i = \sum_{j=1}^{n} a_{i,j} Z_j$, where $Z_1, \ldots, Z_n$ are i.i.d. standard Gaussian random variables. Then the n-dimensional random variable $X = (X_1, \ldots, X_n)^T$ has a joint Gaussian distribution with covariance Σ.*

A famous theorem of Kolmogorov [3, page 382] shows how to construct a stochastic process with a given joint distribution of $(X_{t_1}, \ldots, X_{t_n})$ for any finite set of points $\{t_1, \ldots, t_n\} \subseteq T$. This is called the Kolmogorov extension theorem. Since our focus is on the RKHS aspects of the theory, we will only show that, given a covariance function, there exists a stochastic process with the given covariance. It is important to realize that the Kolmogorov extension theorem is a much more powerful result. However, in the case that the joint distributions of $(X_{t_1}, \ldots, X_{t_n})$ are Gaussian we know that the covariance determines the distribution. Therefore, for Gaussian processes, the following result is equivalent to Kolmogorov's result.

In order to construct a process with a given covariance, we need to use the fact that every kernel K can be written in the form $K(x, y) = \sum_{i \in I} e_i(x)\overline{e_i(y)}$, where e_i is any orthonormal basis or, more generally, a Parseval frame for $\mathcal{H}(K)$, and I is some index set.

Theorem 12.11. *Let K be a kernel function on a set T and let $\{e_i : i \in I\}$ be a Parseval frame for $\mathcal{H}(K)$. Let $\{Z_i : i \in I\}$ be uncorrelated random variables with mean 0 and variance 1 on some probability space Ω. Then $X_t = \sum_{i \in I} e_i(t)Z_i$ converges in $L^2(\Omega)$. If for each t we choose a representative function X_t, then X is a stochastic process with covariance K.*

In addition, if the Z_i are i.i.d. standard Gaussian random variables, then X_t is a Gaussian process.

Proof. Note that the for each t, $\{e_i(t) : i \in I\}$ is square-summable and, hence, is nonzero for at most countably many i. Since the Z_i is orthonormal in $L^2(\Omega)$ it follows that the series converges in $L^2(\Omega)$ and that $\mathbb{E}[X_t] = 0$.

Also

$$\mathrm{Cov}(X_s, X_t) = \mathbb{E}[X_s X_t] - \sum_{i,j \in I} e_i(s)\overline{e_j(t)}\langle Z_i, Z_j\rangle - \sum_{i \in I} e_i(s)\overline{e_i(t)} = K(s, t),$$

so the stochastic process has the desired covariance.

Given a finite set $F \subset I$ set $X_t^F = \sum_{i \in F} e_i(t)Z_i$. If the Z_i are i.i.d. standard Gaussian, then by Lemma 12.10, given any $t_1, \ldots, t_n$ the random variables $X_{t_1}^F, \ldots, X_{t_n}^F$ are jointly Gaussian. It follows, since limits of Gaussians are Gaussian, that the X_t has a Gaussian distribution. $\square$

We now focus on some simple examples of kernel functions and stochastic processes.

Example 12.12. Consider the kernel function $K : T \times T \to \mathbb{R}$ given by $K(t, t) = \sigma^2$, $K(s, t) = 0$ for $s \neq t$. If assume that Z_i are i.i.d. Gaussian in the previous theorem, then the associated process is called the $\boxed{\text{WHITE NOISE}}$ process. The RKHS associated with this kernel is $\ell^2(T)$, where T is viewed as a discrete set.

In general, our kernel functions will exhibit a certain amount of regularity. For instance, we will often deal with continuous kernels defined on compact subsets of $\mathbb{R}^d$ (i.e. Mercer kernels).

In our next example we re-examine the kernel $K(s,t) = \min\{s,t\}$, defined on the interval $T = [0, \tau]$. We will see that the functions $\mathcal{H}(K)$ are not only continuous (see Theorem 2.17) but have bounded variation. This fact is rather interesting, since K is the covariance for the $\boxed{\text{WIENER PROCESS}}$, (also called Brownian motion). The Wiener process has continuous sample paths, of unbounded variation. This illustrates an important point. Although there is a connection between stochastic processes and RKHS, it is not the case that properties of functions in $\mathcal{H}(K)$ are reflected in the sample paths of the process. This should come as no surprise, since processes with very different distributions can have the same covariance kernel.

Example 12.13. Let $T = [0, +\infty)$ and let $K(s,t) = \min(s,t)$, which appeared in the example of the Volterra integral operator. We recall that the space $\mathcal{H}(K)$ is the set of functions that have derivatives in $L^2(T)$, with the norm given by $\|f\|_{\mathcal{H}(K)} = \left(\int_0^{+\infty} |f'(t)|^2 dt\right)^{1/2} = \|f'\|_{L^2(T)}$. This means that every function in $\mathcal{H}(K)$ is of the form $f(t) = \int_0^t g(s)\,ds$, where $g \in L^2(T)$. In this case we write $f' = g$.

Note that every function in $\mathcal{H}(K)$ is continuous, since the kernel function is continuous. However much more than this is true. Every function in $\mathcal{H}(K)$ is of bounded variation. Let us pause to recall some basic terminology related to the variation of a function. Given a function $f : [a, b] \to \mathbb{R}$ and a partition $\mathcal{P} = [t_0, t_1, \ldots, t_{n-1}, t_n]$ of the interval $[a, b]$, we define

$$V(f, \mathcal{P}) := \sum_{i=0}^{n-1} |f(t_{i+1}) - f(t_i)|.$$

We define the total variation of the function f to be

$$TV(f) = \sup_{\mathcal{P}} V(f, \mathcal{P}),$$

where the supremum is over all partitions of the interval $[a, b]$. A function f is said to be of bounded variation on $[a, b]$ if $TV(f)$ is finite.

Theorem 12.14. *Let $K : [0, \tau] \times [0, \tau] \to \mathbb{R}$ be given by $K(s, t) = \min\{s, t\}$.*
Then every function $f \in \mathcal{H}(K)$ is of bounded variation on $[0, \tau]$. Furthermore,
$TV(f) = \|f'\|_{L^1}$.

Proof. We know that $f \in \mathcal{H}(K)$ if and only if there exists a function $\phi \in L^2[0, \tau]$ such that $f(x) = \int_0^x \phi(t)\, dt$, and we denote $f' = \phi$. Furthermore, we have

$$\int_0^\tau |\phi(t)|\, dt \le \left(|\phi(t)|^2\, dt\right)^{1/2} \left(\int_0^\tau dt\right)^{1/2} = \|\phi\|_2 \sqrt{\tau}.$$

Hence, $\phi \in L^1$ and $\|\phi\|_{L^1} \le \sqrt{\tau}\, \|\phi\|_{L^2}$.

Now let $\mathcal{P} = [t_0, t_1, \cdots, t_n]$ be a partition of the interval $[0, \tau]$. We have

$$V(f, \mathcal{P}) = \sum_{i=0}^{n-1} |f(t_{i+1}) - f(t_i)| = \sum_{i=0}^{n-1} \left| \int_{t_i}^{t_{i+1}} \phi(s)\, ds \right|$$

$$\le \sum_{i=0}^{n-1} \int_{t_i}^{t_{i+1}} |\phi(s)|\, ds = \int_0^\tau |\phi(s)|\, ds = \|\phi\|_{L^1}.$$

This establishes the fact that $TV(f) \le \|f'\|_{L^1}$.

We prove the reverse inequality by first checking the result for functions in the span of the kernel functions. We then obtain the equality for all functions in $\mathcal{H}(K)$ by a density argument.

To this end, let $f = \sum_{i=1}^m a_i k_{x_i}$ where $0 < x_1 < \cdots < x_m \le \tau$. Note that $k_0 \equiv 0$, and so we may assume that $x_1 > 0$.

For a kernel function k_x we have,

$$k_x'(t) = \begin{cases} 1 & t < x \\ 0 & x \le t \le \tau \end{cases}.$$

Hence,

$$f'(t) = \begin{cases} \sum_{i=1}^m a_i & t < x_1 \\ \sum_{i=2}^m a_i & x_1 \le t < x_2 \\ \vdots & \\ a_m & x_{m-1} \le t < x_m \\ 0 & x_m \le t \le \tau \end{cases}.$$

If we let $b_j = \sum_{i=j}^m a_i$, then

$$\int_0^\tau |f'(t)|\, dt = x_1 |b_1| + (x_2 - x_1) |b_2| + \cdots + (x_m - x_{m-1}) |b_m|.$$

On the other hand, consider the partition $\mathcal{P} = \{0, x_1, x_2, \ldots, x_m, \tau\}$. We have $f(0) = 0$ and $f(\tau) = \sum_{i=1}^{m} a_j x_j$. In addition, for $1 \leq j \leq m$ we have

$$f(x_{j+1}) - f(x_j) = \sum_{i=1}^{m} a_i (K(x_i, x_{j+1}) - K(x_i, x_j)).$$

For $i \leq j$ the expression $K(x_i, x_{j+1}) - K(x_i, x_j) = x_i - x_i = 0$. For $i \geq j+1$ we get $x_{j+1} - x_j$. Hence,

$$f(x_{j+1}) - f(x_j) = \sum_{i=j+1}^{m} a_i (x_{j+1} - x_j) = (x_{j+1} - x_j) b_{j+1}.$$

In addition, $f(x_1) - f(0) = f(x_1) = x_1 b_1$ and $f(\tau) - f(x_m) = 0$. Hence, $V(f, \mathcal{P}) = x_1 |b_1| + \cdots + (x_m - x_{m-1}) |b_m|$. It follows that $TV(f) = \|f'\|_{L^1}$.

Now suppose that $f \in \mathcal{H}(K)$ and let g be a function in the span of the kernels such that $\|f - g\|_2 < \epsilon/\sqrt{\tau}$. Let $\mathcal{P}$ be a partition of $[0, \tau]$. The absolute value of the difference between the variations over $\mathcal{P}$ is given by

$$
\begin{aligned}
|V(f, \mathcal{P}) - V(g, \mathcal{P})| &= \left| \sum_{i=0}^{n-1} \left| \int_{t_i}^{t_{i+1}} f'(t)\,dt \right| - \left| \int_{t_i}^{t_{i+1}} g'(t)\,dt \right| \right| \\
&\leq \sum_{i=0}^{n-1} \int_{t_i}^{t_{i+1}} |f'(t) - g'(t)|\,dt \\
&\leq \int_0^{\tau} |f'(t) - g'(t)|\,dt = \|f' - g'\|_{L^1} \\
&\leq \sqrt{\tau} \|f - g\|_{\mathcal{H}(K)} < \epsilon.
\end{aligned}
$$

Hence, $V(f, \mathcal{P}) \leq V(g, \mathcal{P}) + \epsilon$ for all partitions. It follows that $TV(f) \leq TV(g) + \epsilon$ for all ϵ. In particular $TV(f)$ is finite. Furthermore, $TV : \mathcal{H}(K) \to \mathbb{R}$ is continuous. Since $\|f'\|_{L^1} \leq \sqrt{\tau} \|f\|_{\mathcal{H}(K)}$ it follows that $TV(f) = \int_0^{\tau} |f'(t)|\,dt$. □

A function $f : [a, b] \to \mathbb{R}$ is of bounded quadratic variation if there exists a constant B such that, for any partition $\mathcal{P} = \{t_0, \ldots, t_n\}$ of the interval $[a, b]$, we have, $\sum_{i=0}^{n-1} |f(t_{i+1}) - f(t_i)|^2 \leq B$. The smallest such constant is called the quadratic variation of f. We require only RKHS methods to show that every function in $\mathcal{H}(K)$ is of bounded quadratic variation.

Proposition 12.15. *Let $K = \min(s, t)$ on the interval $[0, \tau]$, then every function $f \in \mathcal{H}(K)$ is of bounded quadratic variation. The quadratic variation of f is given by $\|f\|_{\mathcal{H}(K)}^2 \tau$.*

We leave the proof as an exercise.

We now present a second example of a stochastic process. This example is finite-dimensional.

Example 12.16. Let $\phi_1, \ldots, \phi_m$ be m functions defined on some set T. Consider the functions of the form $f = \sum_{i=1}^{m} w_i \phi_i$.

We will first endow the set of functions $\sum_{i=1}^{m} w_i \phi_i$ with the structure of an RKHS. To do this, let $K(s, t) = \sum_{i=1}^{n} \phi_i(s) \phi_i(t)$. We have seen in 5.3 that K is a kernel function and that $\mathcal{H}(K)$ is finite-dimensional. Note that ϕ_i is a Parseval frame for $\mathcal{H}(K)$. Hence, if $f = \sum_{i=1}^{m} w_i \phi_i$, then $\|f\|^2_{\mathcal{H}(K)} = \sum_{i=1}^{n} |w_i|^2 = \|w\|^2$.

We now want to construct a stochastic process with covariance K. Using theorem 12.11 we know that if $W_1, \ldots, W_m$ are i.i.d. Gaussians, then $F_t = \sum_{i=1}^{m} W_i \phi_i$ is a Gaussian process with covariance K. Let Ω be the underlying probability space for $W_1, \ldots, W_m$. A sample path of F is a function of the form $f = \sum_{i=1}^{m} W_i(\omega) \phi_i$. This function is a random element of the RKHS $\mathcal{H}(K)$.

The stochastic process F induces a probability distribution on the space $\mathcal{H}(K)$. To see this, consider the map $u : \Omega \to \mathcal{H}$ given by $u(\omega) = \sum_{i=1}^{m} w_i(\omega) \phi_i$. The function u is a random variable with values in the Hilbert space $\mathcal{H}(K)$. For each Borel subset of $E \subseteq \mathcal{H}(K)$ we can define the induced measure $\mathbb{P}^*(E) = \mathbb{P}(u^{-1}(E))$.

Let us compute the mean and covariance of the stochastic process F. We have

$$m_F(t) = \mathbb{E}\left(\sum_{i=1}^{n} W_i \phi_i(t)\right) = 0.$$

To compute the covariance we use the fact that the W_i are independent. In particular, $\mathbb{E}[W_i W_j] = 0$ whenever $i \neq j$. Hence,

$$\text{Cov}(F(s), F(t)) = \sum_{i,j=1}^{m} \mathbb{E}[W_i W_j] \phi_i(s) \phi_j(t)$$

$$= \sum_{i=1}^{m} \mathbb{E}[W_i^2] \phi_i(s) \phi_i(t) = K(s, t).$$

This last example is the protypical example of a Gaussian process (albeit in finite dimensions). The reader who wants to learn more about the connections between Gaussian processes and RKHS should read Rasmussen and Williams [14].

12.4 Karhunen–Loève theorem

In this section we prove the Karhunen–Loève theorem, which allows us to decompose a L^2-stochastic process in terms of a special set of uncorrelated random variables. In Theorem 12.11 we saw how to construct a stochastic process with any given covariance function $K(s, t)$ by choosing an orthonormal basis or Parseval frame for $\mathcal{H}(K)$ and using any family of uncorrelated random variables. The Karhunen–Loève theorem is a natural counterpart of this construction. Given the stochastic process, one would like to express it as such a sum of uncorrelated random variables using uncorrelated random variables that are "natural" to the process.

If we assume that the kernel K is a Mercer kernel, then there is a natural choice of basis obtained from Mercer's theorem 11.11. In this case the random coefficients in the expansion can be computed from the process X. The proof assumes that the reader is familiar with the theory of integration of Banach space-valued functions.

Theorem 12.17 (Karhunen–Loève theorem). *Let $X : T \times \Omega \to \mathbb{R}$ be a stochastic process on a compact subset $T \subseteq \mathbb{R}^d$ such that the covariance function K is a Mercer kernel on T. Let $T_K : L^2(T) \to L^2(T)$, be the integral operator for Lebesgue measure with symbol K and let e_k be the continuous eigenfunction of T_K corresponding to the eigenvalue $\lambda_k > 0$ given by Mercer's theorem. Set*

$$Z_k = \frac{1}{\lambda_k} \int_T X_t e_k(t) \, dt.$$

Then $\{Z_k : k = 1, 2, \dots\}$ is a sequence of pairwise uncorrelated random variables in $L^2(\Omega)$, with mean 0 and variance 1. Furthermore,

$$X_t = \sum_{k=1}^{\infty} Z_k e_k(t), \tag{12.1}$$

where the series on the right-hand side of Eq. (12.1) converges in $L^2(\Omega)$. If the process is Gaussian, then Z_k is Gaussian for all $k \in \mathbb{N}$.

Proof. To prove this theorem we appeal to Mercer's theorem and express $K(s, t) = \sum_{k=1}^{\infty} e_k(s) \overline{e_k(t)}$, where the function e_k is an orthonormal basis for $\mathcal{H}$, and $T_K(e_k) = \lambda_k e_k$ for some positive eigenvalue $\lambda_k > 0$. It is also the case that $\|e_k\|_{L^2(T)} = \sqrt{\lambda_k}$, and the functions e_k are pairwise orthogonal in $L^2(T)$.

Consider the function $t \mapsto X_t$. We have,

$$\|X_t - X_s\|^2 = K(s,s) + K(t,t) - 2K(s,t),$$

and so the map $t \mapsto X_t$ is continuous from T to $L^2(\Omega)$. Since e_k is also continuous, we see that $t \to e_k(t)X_t$ is a continuous, and hence uniformly continuous, function from the compact set T into the Banach space $L^2(\Omega)$. In this case, the Bochner integral

$$Z_k = \frac{1}{\lambda_k} \int_T X_t e_k(t) \, dt \in L^2(\Omega)$$

is well-defined. When T is an interval or rectangle, this integral is just a limit of Riemann sums.

We now show that the sequence $(Z_k)_{k \in \mathbb{N}}$ is uncorrelated, and then compute the variance of Z_k. We have

$$
\begin{aligned}
\langle Z_n, Z_m \rangle &= \frac{1}{\lambda_n \lambda_m} \left\langle \int_T X_t e_n(t) \, dt, \int_T X_s e_m(s) \, ds \right\rangle \\
&= \frac{1}{\lambda_n \lambda_m} \iint_{T \times T} \langle X_t, X_s \rangle \, e_n(t) e_n(s) \, ds \, dt \\
&= \frac{1}{\lambda_n \lambda_m} \iint_{T \times T} K(s,t) e_n(t) e_n(s) \, ds \, dt \\
&= \frac{1}{\lambda_n \lambda_m} \langle T_K e_n, T_K e_m \rangle = \frac{1}{\lambda_n \lambda_m} (\lambda_n \lambda_m \langle e_n, e_m \rangle) = \delta_{m,n}.
\end{aligned}
$$

In addition we see that $\int_\Omega Z_k = \int_\Omega \int_T X_t e_k(t) \, dt = \int_T e_k(t) \int_\Omega X_t \, d\omega = 0$, since we assumed that $\mathbb{E}[X_t] = 0$.

We now prove that Z_k is Gaussian whenever X_t is a Gaussian process. To see this note that Z_k can be approximated by sums of the form $\sum_{i=1}^n \mu(E_i) X_{t_i} e_k(t_i)$. Since a sum of Gaussian random variables is Gaussian, as is a limit, we see that Z_k is Gaussian. Using the fact that uncorrelated Gaussian random variables are independent we get that Z_k is an independent sequence of random variables.

Finally we need to show that the series $\sum_{k=1}^\infty Z_k e_k(t)$ converges to the process X_t. First we compute

$$
\begin{aligned}
\langle X_t, Z_k \rangle &= \lambda_k^{-1} \left\langle X_t, \int_T X_s e_k(s) \, ds \right\rangle \\
&= \lambda_k^{-1} \int_T \langle X_t, X_s \rangle \, e_k(s) \, ds \\
&= \lambda_k^{-1} \int_T K(s,t) e_k(s) \, ds = e_k(t).
\end{aligned}
$$

We have

$$\left\| X_t - \sum_{k=1}^n Z_k e_k(t) \right\|_{L^2(\Omega)}^2 = \|X_t\|^2 - 2\sum_{k=1}^n e_k(t) \langle X_t, Z_k \rangle$$

$$+ \sum_{k,l=1}^n \langle Z_k, Z_l \rangle e_k(t) e_l(t)$$

$$= K(t,t) - 2\sum_{k=1}^n e_k(t)^2 + \sum_{k=1}^n e_k(t)^2$$

$$= K(t,t) - \sum_{k=1}^n e_k(t)^2.$$

This last sum goes to 0 as $n \to \infty$. In particular, the convergence is uniform in t by Proposition 11.8. $\square$

12.5 Exercises

Exercise 12.1. This problem focuses on the function

$$K(s,t) = \min\{s,t\} - st,$$

the associated RKHS and Gaussian process.

1. Show that K is a kernel on $[0,1]$.
2. Let W_t be the process with covariance $\min\{s,t\}$, i.e. the standard Wiener process. Show that $B_t = W_t - tW_0$ has the above covariance function.
3. Using the functions $\phi(s,x) = \mathbf{1}_{[0,s]} - s$ show that

$$K(s,t) = \int_0^1 \phi(s,x)\phi(t,x)\,dx.$$

4. Use this to show that $\mathcal{H}(K)$ is the space of functions that have weak L^2-derivatives and satisfy $f(0) = f(1) = 0$, with the norm given by $\|f\|_{\mathcal{H}(K)} = \|f'\|_{L^2[0,1]}$.
5. Let W_t be the standard Wiener process; show that K is also the kernel function for the process $B_t = (W_t | W_1 = 0)$.

The stochastic processes B_t are called the Brownian bridge, or pinned Wiener process.

Exercise 12.2. An Ornstein–Uhlenbeck process is a Gaussian process with covariance function $K(s,t) = \mathrm{Cov}(X_s, X_t) = e^{-|s-t|}$ for $s, t \in \mathbb{R}$. Show directly that this is a kernel function on $\mathbb{R}$. Solve the reconstruction problem for K, i.e. give a description of the functions in the space $\mathcal{H}(K)$.

Exercise 12.3. Prove Theorem 12.4.

Exercise 12.4. Prove Propostion 12.10.

Exercise 12.5. Prove Proposition 12.15.

Bibliography

[1] Jim Agler and John E. McCarthy, *Pick interpolation and Hilbert function spaces*, Graduate Studies in Mathematics, vol. 44, American Mathematical Society, Providence, Rhode Island, 2002.

[2] N. Aronszajn, *Theory of reproducing kernels*, Trans. Amer. Math. Soc. **68** (1950), 337–404. MR0051437 (14,479c)

[3] R. F. Bass, *Stochastic processes*, Cambridge Series in Statistical and Probabilistic Mathematics, vol. 33, Cambridge University Press, Cambridge, UK, 2011.

[4] S. Bergman, *Ueber die Entwicklung der harmonischen Funktionen der Ebene und des Raumes nach Orthogonalfunktionen*, Math. Ann. **86**, (1922).

[5] Louis de Branges and James Rovnyak, *Square summable power series*, Holt, Rinehart and Winston, New York, 1966. MR0215065 (35 #5909)

[6] Louis de Branges, *Hilbert spaces of entire functions*, Prentice-Hall Inc., Englewood Cliffs, N.J., 1968. MR0229011 (37 #4590)

[7] John B. Conway, *A course in functional analysis*, 2nd ed., Graduate Texts in Mathematics, vol. 96, Springer-Verlag, New York, 1990. MR1070713 (91e:46001)

[8] Ronald G. Douglas and Vern I. Paulsen, *Hilbert modules over function algebras*, Pitman Research Notes in Mathematics, vol. 217, Longman Scientific, 1989.

[9] Gerald B. Folland, *Real analysis: modern techniques and their applications*, John Wiley & Sons, Inc., New York, 1999.

[10] E. H. Moore, *General analysis. 2*, Vol. 1, 1939.

[11] Sheldon Ross, *A first course in probability*, 9th ed., Pearson Education Limited, Harlow, Esex, England, 2012.

[12] Donald Sarason, *Complex function theory*, American Mathematical Society, Providence, Rhode Island, 2007.

[13] Elias M. Stein and Rami Shakarchi, *Real analysis: measure theory, integration, and Hilbert spaces*, Princeton Lectures in Analysis III, Princeton University Press, Princeton, New Jersey, 2005.

[14] Christopher K. I. Williams and Carl Edward Rasmussen, *Gaussian processes for machine learning*, MIT Press, 2006.

Index